The Policy Analysis Matrix
for Agricultural Development

THE
Policy Analysis Matrix
FOR
Agricultural Development

Eric A. Monke *and*
Scott R. Pearson

Cornell University Press

ITHACA AND LONDON

First published 1989 by Cornell University Press.

International Standard Book Number (cloth) 0–8014–1953–0
International Standard Book Number (paper) 0–8014–9551–2
Library of Congress Catalog Card Number 88–47938
Printed in the United States of America
Librarians: Library of Congress cataloging information
appears on the last page of the book.

The paper in this book is acid-free and meets the guidelines for
permanence and durability of the Committee on Production Guidelines
for Book Longevity of the Council on Library Resources.

For Kim and Sandra

Contents

Preface

POLICIES FOR AGRICULTURE consist of government decisions that
influence the level and stability of output and input prices, public investments affecting agricultural revenues and costs, and the allocation of
research funds to improve farming and processing technologies. Some
of these policies are specific to agriculture–fertilizer subsidies and tariffs
on wheat imports, for example–whereas others, such as fiscal and
exchange-rate policy, affect all sectors of the economy. How can applied economists interpret the effects on agriculture of such a variety of
policies? This book presents an organizational framework–the policy
analysis matrix (PAM)–to serve this need.

The PAM allows applied economists to analyze policies in terms of
their impact on commodity systems–representative chains of farming,
marketing, and processing activities that together produce a marketable
product such as wheat flour or milled rice. Choice of a particular
commodity system as the focal point of analysis is arbitrary. Alternative
economic criteria, such as the provision of basic needs or employment
and wages in the labor market, could provide equally valid perspectives
for measuring the impacts of policy on agriculture. But in most cases,
policy-makers are vitally interested in the effects of policies on commodities, even when the policies are not formulated to further specific
commodity objectives. Both the formulation of a theoretical rationale
and the empirical implementation of policy analysis are easier when
commodities serve as the organizational framework. The impact of
policies on other elements, such as employment and wages, can then be
pursued through the aggregation of commodity system results.

The PAM is a compromise between the desire for a theoretical model to describe the economy in exacting detail and the need for insightful policy analysis that operates within the inevitable constraints of time and data availability. The theoretical basis for PAM is the simple general equilibrium model of international trade rather than some social welfare function, and the matrix focuses attention on the identification of efficient patterns of production and prices. Nonefficiency objectives are then considered as potential justifications for policies that support inefficient production systems. This restricted approach to the identification of the optimum policy set is more helpful to informed policy debate than are analyses based on a priori inferences about the "proper" roles for efficiency and nonefficiency objectives of agricultural policy. Policy debate most often arises because comparisons of the importance of various objectives are neither predetermined nor quantifiable.

For empirical application of the PAM, emphasis is placed on budgets for costs and returns, which are chosen to represent commodity systems for different regions, types of farms, and technologies. Budgets easily accommodate the effects of direct policy interventions that alter the commodity and factor prices of the commodity system. However, the indirect impacts of policy distortions are less easily quantified. Indirect policy influences include the effect of all price distortions on the exchange rate, which in turn alters the prices of output and of some inputs of the commodity system; the effects of output price distortions on the prices of domestic factors (labor, capital, and land); and the effects of input substitution on factor prices. In the modification of private costs and returns to approximate social values, these indirect effects are ignored or are recognized only in a very approximate manner. In principle, econometric models could be used to calculate the impacts of these indirect effects. But in practice, information and resource constraints have often meant that such estimates are unreliable or ad hoc.

This book focuses on the agriculture of developing countries. No logical or practical reasons preclude application of the method to the agricultural sector of any country. The options for policy intervention are finite, and the relationships between policies and incentive effects are largely independent of the level of economic development. But sectoral objectives and policy choices often differ substantially between developing and developed countries. For example, high-income countries give prominence to the maintenance of producers' income and to environmental externalities. Consequently, a book about policy analysis of agriculture in developed countries would discuss alternative price and income subsidy programs and the evaluation of nonmarket goods in a

more elaborate manner than is provided here. Moreover, data availability and analytical resources are more ample in developed countries,
allowing more discussion of alternative methods of empirical analysis.
But our interests and previous work have concentrated almost exclusively on developing countries, and the book has been written with
those countries in mind.

We have tried to address the needs of two groups of development
economists—users (academics, students, and policy analysts who can
use the PAM as a conceptual approach to policy analysis) and doers
(practitioners interested in applying the approach). We believe the PAM
is more than a technique for evaluating policy options; it also provides a
way of thinking about policy. The first two chapters of the book develop
this line of argument. Chapters 3 through 5 deal with the rationales for
agricultural policy interventions and provide a heuristic discussion of
the impacts of various policies—commodity, factor, and macroeconomic—on an agricultural system. Chapters 6 and 7 use the general
equilibrium model of international trade to show how divergences
(distorting policies and market failures) affect the values of outputs and
inputs associated with an agricultural system.

The fourth part of the book is a discussion of the empirical procedures usually followed in the construction of PAMs. Chapters 8
through 11 are concerned with the collection of data and the estimation
of PAMs; chapter 12 focuses on analysis and presentation of results.
This material might seem essential only for those contemplating a piece
of applied research, but users of the PAM approach will also benefit
from understanding more about the practice of policy analysis. Therefore, we have chosen to make this material available to all readers, users
as well as doers. The book concludes with a summary chapter on the
practice of agricultural policy analysis.

We owe thanks for the ideas, insights, and comments of a great many
colleagues and students. We particularly thank, without implication,
Michele de Benedictis, Roger Fox, Charles Humphreys, Timothy Josling, William Masters, and Karen Parker. Barbara O'Leary prepared the
original manuscript and several revisions, Linda Phipps provided the
graphics, and Claudia Smith typed several early drafts of chapters; we
are grateful for their skill and patience.

Eric A. Monke
Scott R. Pearson

Tucson, Arizona
Stanford, California

PART I

INTRODUCTION

CHAPTER 1

The Role of
Agricultural Policy Analysis

INTEREST IN THE ANALYSIS of agricultural policy is a relatively recent phenomenon. Before the mid-1960s, industrialization was seen as the key to economic development in most developing countries. Accordingly, government resources and policies were focused on the promotion of industry, and the agricultural sector was thought of primarily as a pool of resources for the development of the nonagricultural sector. Capital for new industrial investment would be obtained from taxes on the rural population or on agricultural output; labor requirements would be met by the removal of labor from the agricultural sector. Because much of the agricultural labor force was thought to be unproductive, food production would not decline.

Problems soon arose with this industry-first strategy. In many developing countries, the small size of domestic markets forced new industries to compete in international markets. But competing firms in other countries were often more efficient. To sustain the viability of firms, either domestic consumers had to pay prices for industrial output that were higher than world prices or governments had to subsidize production costs to maintain international competitiveness. Often, both policies were used. The result was an increasing burden on the budgets of consumers and governments. A second problem with the industry-first strategy was the absence of surplus resources in agriculture. In most countries, industry had to lure labor away from productive agriculture, and agricultural production declined as industrial production grew. Foreign exchange had to be directed increasingly toward imports of food rather than imports of the inputs essential for industrial development. Government revenues from agricultural taxation declined as well.

3

Finally, the industry-first strategy encouraged a deterioration in the rural-urban income distribution. Income increases were concentrated in urban areas, and benefits for workers in the agricultural sector were limited largely to the workers who were able to emigrate successfully to urban areas. Because agricultural incomes did not increase, rural demands for industrial sector outputs remained small.

In the past two decades, new development strategies have emerged. Agriculture has been placed at the forefront, and industrial development has become a complement to agricultural growth. Expansion of agricultural production is seen as leading to increases in farm income that fuel the demands for industrial sector outputs. Initially, industrial production is dominated by industries that produce agricultural inputs, low-cost consumer goods, and construction and transportation services. More complex industries develop as the supply of entrepreneurial and managerial talent increases, sustained by public investment in education and infrastructure. In contrast to the earlier approaches, the industrial-growth strategy is determined largely by domestic demand. International markets can still provide opportunities for growth, but these markets are exploited only as competitive industries emerge. Processing of agricultural and other labor-intensive products dominates potential export industries.

In the current strategy, agricultural policy is a critical element in determining the rate and pattern of economic growth. One set of policies—investment in education, health and sanitary facilities, and transportation infrastructure—has a broad impact on agricultural sector productivity. In general, economists, policy-makers, and development institutions have reached a consensus on the importance of these investments. A second set of policies affects particular agricultural commodities or techniques of production. These commodity-specific policies include taxes, subsidies, and quantitative controls on particular outputs and inputs, and policies that affect the macroprices (interest rates, wage rates, and exchange rates). For this set of policies, little consensus has emerged on appropriate levels of use. Analysis of this second category of policies is the principal concern of this book.

Rationales for Intervention

One reason that governments impose policies on their agricultural sector is the belief that intervention can accelerate the rate of income growth. Investment policies—the provision of public goods, such as the research and development of new technologies and infrastructural development (roads, schools, health facilities)—are examples of public sector interventions essential for increased economic activity. Sometimes, these investments will not be made by the private sector. Private investors may be unable to capture the full benefit from investment in public goods because it is impossible or too costly to exclude those who do not pay for services created. In other instances, consumption by one consumer does not reduce the availability of the good or service for others. Consumers therefore avoid declaring their willingness to pay for the good or service, and a market does not form. Finally, capital requirements of the investment might exceed the private sector's capacity to mobilize necessary financial resources. For most of these investments, the public sector has the potential to recover the costs of intervention through user fees or through taxation of the commodities or the regional populations that benefit from the investment.

The correction of market failures represents a second rationale for government intervention in the agricultural sector. If market imperfections are present, the prices of goods or services will not reflect their true scarcity values because the private sector is unable to develop the institutions necessary for efficient market functioning. Rural credit markets, for example, might be hampered by a lack of information on alternative lending and borrowing opportunities in other regions, or by the absence of formal lending institutions that can mobilize savings. Market power is another example of a market failure; private sector suppliers (or consumers) are able to influence prices because their numbers are small and because buyers (or sellers) have no other market outlets. These conditions are asserted to prevail often in factor markets (those for labor, credit, and land) and sometimes in remote rural commodity markets.

Another type of market failure arises because of externalities—costs or benefits from production activities that are not fully reflected in market incentives. Soil erosion, environmental pollution, and overutilization of common property resources are common externalities. Some form of government intervention—a tax, subsidy, or regulatory control—is justified so that user costs (or returns) will reflect fully the effects of the externality. The value of an externality is often difficult to quantify, and in many cases subjective judgments must be made as to whether externality effects are significant. These measurement problems, combined with the administrative costs of tax and subsidy policies, cause quantitative or legislative regulations to be widespread policy responses to externalities.

Although policies to correct market failures or to provide public goods can be important, the most common rationale for intervention in developing country agriculture is the promotion of nonefficiency objectives. The establishment of an efficient economy and the maximization of aggregate income are not the only, or necessarily the most important, goals of economic policy. When policy-makers are dissatisfied with the implications of income maximization, policies will be used to alter the economy. In some cases, these interventions will reflect neutral policy-makers acting on a mandate from society. But more often, policies respond to the desires of special interest groups within or outside agriculture.

Income distribution concerns are often at the top of the list of nonefficiency objectives. Food is the most basic of necessities, and low prices of food are considered an important determinant of the welfare level of poor consumers. Staple food prices influence producer income levels as well, and the manipulation of producer prices may generate a more equitable distribution of income in the economy. Income distribution policies will also reflect the influences of rent-seekers—agricultural commodity producers and input suppliers, consumers of food, and industrialists who view changes in agricultural prices as ways to increase profitability in production or to increase purchasing power in consumption. Government policies can benefit target groups through direct regulation of prices—such as tariffs or subsidies on imports—or through policies that provide market power to the target group, such as the designation of monopoly suppliers of particular agricultural products or the allocation of import and export licenses.

Price stabilization is a second common justification for intervention in agriculture. Dependence on the weather causes agricultural production to exhibit a relatively large degree of random variation. When combined with inelastic demand, supply variations can cause market prices to fluctuate substantially from one production cycle to the next. The consequent potential income fluctuations for poor producers and variations in expenditure for poor consumers are often unacceptable to policy-makers. To avoid substantial fluctuations in domestic market prices, many governments establish a set of policies, choosing among international trade controls, storage schemes, price fixing, and rationing. Elements of market failure are also partially responsible for interventions of this type. In production, for example, crop insurance and futures and options markets are institutions that reduce the uncertainty of future prices and income. However, these institutions are usually absent from developing country markets.

National concern over the appropriate role for agriculture in the economy provides a third set of nonefficiency rationales for government intervention. Food security and self-reliance of staple food supplies are commonly held objectives for agricultural policy. For food-importing countries, the attainment of these objectives requires intervention to increase domestic production. This intervention might involve changes in producer prices of outputs and inputs, investment in infrastructure for production or marketing activities, or quantitative restrictions on the production of alternative crops. Agriculture also contributes to government revenue and the maintenance of fiscal balance in the public sector. Income taxes are a relatively unimportant revenue source in most developing countries because informal methods of income payment are prominent. As a consequence, the adminstrative costs of income monitoring and tax collection are often prohibitive, and indirect taxes on commodities are an important source of revenue. Because of its large size, the agricultural sector is usually expected to play a prominent role in the generation of tax revenues.

The relative importance of each justification for intervention in the agricultural sector follows no particular pattern across countries. In part, this variation results from wide disparities in the distribution of political power. The importance and effectiveness of various lobbying groups—domestic producers, consumers, government agencies, and foreign governments and corporations—vary enormously across countries. Consequently, cross-country variations in agricultural sector objectives are large. Differential resource constraints also create cross-country differences in agricultural objectives. Governments have objectives for sectors other than agriculture, which implies that budget constraints are a potential limitation on agricultural sector interventions. Technological limitations also might mean that some objectives cannot be realized at reasonable cost. To some extent, policy-makers can overcome constraints by judiciously selecting policies. Selection of the policy that minimizes budgetary cost allows the furtherance of more objectives than would otherwise be the case. But, ultimately, constraints in most developing countries become binding well before all the objectives of agricultural policy can be realized.

The Evaluation of Policy

Given the importance of nonefficiency objectives, evaluation of the tradeoffs that arise between efficiency and nonefficiency objectives assumes particular interest in policy analysis. Because resources are in limited supply, the achievement of any particular objective will usually come at the expense of reduced activity in some other economic endeavor. The offsetting of market failures is an important exception to this generalization, because these policy interventions liberate resources from less efficient uses and thus increase the total value of economic activity. But in most cases, the attainment of objectives entails economic costs, and the assessment of these tradeoffs can yield insight about the desirability of furthering a particular objective.

A simple graphical description of the tradeoff between efficiency and nonefficiency objectives is provided in Figure 1.1. The curve ADCB portrays the maximum level of production possibilities for a country that produces two commodities, grain and cotton. Producing at world prices leads to a production pattern represented by point C (cotton and grain production are denoted as Q_1^C and Q_1^G, respectively) and to a consumption possibilities frontier, WCZ. By trading at world prices, the country can choose to consume at any point along WCZ. Total income of the country can be measured with respect to either commodity. In terms of grain, total purchasing power is 0W; in terms of cotton, total purchasing power is 0Z.

If the government is dissatisfied with the degree of food self-suffi-

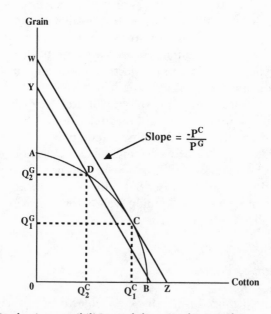

Figure 1.1. Production possibilities and the gains from trade

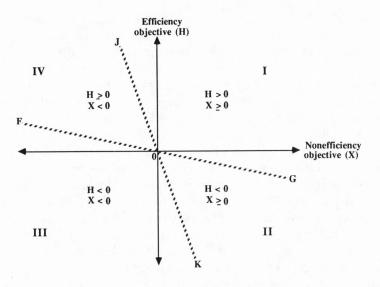

Figure 1.2. Optimal policy choice with multiple objectives

ciency that results from output combination Q_1^C and Q_1^G, it could increase the relative price of grain. Production will then shift to point D. Because the country cannot influence world prices, the slope of the consumption possibilities frontier (WCZ) will not change. It will shift inward to YDB, however, because the frontier must intercept production point D. The country can trade only on the basis of commodities that it has available. Measured in terms of grain, the potential income of the country will fall to 0Y. But under the new policy, a larger share of grain is produced by domestic sources. The difference in total income (0W − 0Y) times the world price of grain equals the efficiency cost of pursuing the nonefficiency objective.

If higher income as well as greater self-sufficiency is desired, the policy-maker is forced to make tradeoffs between objectives. Some compromise must be reached between the desire to maximize consumption possibilities and the interest in increasing domestic food production. Figure 1.2 shows how the tradeoffs between objectives can be analyzed. The y-axis portrays the net addition to potential national income from the commodity system under study. This value, H, is net of all opportunity costs (for resources that can be employed elsewhere in the economy) and thus represents social profit. If the economy is operating at point C of Figure 1.1, each commodity system will show zero social profit. National income is maximized, and input costs exhaust all revenue.

An index of a nonefficiency objective under study is placed on the x-axis. The zero point can be taken as representative of the state of affairs in the absence of policy. For example, if self-sufficiency is the objective, the percentage share of domestic production in domestic consumption can serve as an index measure. Movements along the x-axis rightward from the intersection represent increases in the share of domestic production relative to domestic consumption; movements leftward indicate declines in the share of production.

With the graph, new commodity systems, new technologies, or new policies can then be evaluated in terms of their aggregate potential to increase or decrease the self-sufficiency ratio and to increase or decrease national income. Each commodity system is represented as a point on the graph. If the new commodity system can be located in quadrant I or III, choices for the policy-maker are easy. In quadrant I, no tradeoff exists between objectives. Systems in quadrant I are socially profitable ($H \geq 0$) and contribute positively to the nonefficiency objective ($X \geq 0$). Systems that occupy quadrant III should be discouraged by policy-makers, since those systems decrease national income ($H < 0$) and do not encourage the nonefficiency objective ($X < 0$).

Quadrants II and IV are the areas of difficult policy choice, because they correspond to situations of tradeoffs between objectives. In quadrant II, the new situation encourages the attainment of nonefficiency objectives ($X \geq 0$), but only at a cost in potential national income ($H < 0$). Because $H < 0$, policy-makers must enact policies that subsidize the system; otherwise, production will not be undertaken by the private sector. In the grain-cotton example, this subsidy was effected by an increase in the price of grain. In quadrant IV, a socially efficient system ($H \geq 0$) contributes negatively to the nonefficiency objective but positively to national income.

Evaluation of the systems in quadrants II and IV requires knowledge of the policy-makers' preference locus—the set of points describing the policy-makers' willingness to trade off one objective for the other. Points on this locus represent the amount of income gain needed to compensate for a given reduction in the nonefficiency objective (or, conversely, the amount of gain in the nonefficiency objective that will compensate for a given loss in income). Policy-makers who place a premium on total national income (efficiency) will have a slightly sloped locus (such as F0G); those with relatively strong concerns for food self-sufficiency (nonefficiency) will have a steeply sloped locus (such as J0K).

Two types of policy interventions are needed. Systems represented by points to the right of the preference locus should be encouraged. Systems that are socially unprofitable but that contribute sufficiently to nonefficiency objectives need to be encouraged by policy so that private profitability becomes positive. If J0K represents the preference locus, systems located in the triangular region between 0K and the positive x-axis would merit assistance. Points to the left of the locus indicate systems that create unacceptable tradeoffs between alternative objectives. Policy-makers should discourage systems that are socially profitable but that create too negative an impact on nonefficiency objectives. Systems located in the triangular region between J0 and the negative x-axis warrant taxes so that private profitability will become negative.

In these circumstances, policy analysis appears a straightforward exercise. The analyst need only evaluate profitability and nonefficiency effects associated with commodity systems, and appropriate policy interventions are identified. Comparisons among alternative systems allow the policy analyst to identify least-cost ways of achieving nonefficiency objectives. Systems that allow attainment of the nonefficiency objective at lesser cost (or greater gain) in efficiency terms are always preferable.

The difficulty for policy analysis, however, lies in attempting to identify the exact location of the preference locus. In some cases, observation of policy actions might help to define the locus. For example, if governments vigorously tax nonfood crops and subsidize food crops, the preference locus might be something like J0K. But most situations are unlikely to be so well defined. The individuals who make policy, and their opinions about the appropriateness of various objectives, change frequently. Nor will societal preferences be uniform. As a result, a consensus on appropriate and inappropriate policy actions will not be stable; in many cases, a consensus will not even exist.

Identification of the appropriate tradeoffs between efficiency and nonefficiency is further complicated because governments hold many nonefficiency objectives and impose many policies simultaneously. Commodity policies (taxes, subsidies, and quantitative controls on commodities), macroprice policies (wage rate, interest rate, land rental rate, and exchange rate), and macroeconomic policies (fiscal and monetary management) will exert simultaneous impacts on a commodity system. The net impact of government policy—and hence the true importance of a particular objective—can be assessed only through aggregation of these incentive effects. Expansion of staple food production might be a stated objective for the agricultural sector, for example. But if producers are subjected to high net taxes on production, some skepticism is justified regarding the priority of policy-makers for this objective.

The Role of Quantitative Policy Analysis

Even if the appropriate tradeoffs between efficiency and nonefficiency objectives are not known, quantitative analysis of the economic impacts of policies retains immense importance. But rather than inform the government as to the appropriate actions it should be taking (or not taking), policy analysts provide fuel for the on-going debate between those who wish to change policies and those who wish to maintain them. Few, if any, policies are immutable, and disaggregated information about efficiency and nonefficiency effects of policy allows policy-makers to form opinions about "good" and "bad" policies on an individual basis. Appropriate policy then emerges as a result of negotiation among those with potential to influence policy.

Quantitative policy analysis also plays a dynamic role in the policy-making process by ensuring that agricultural sector objectives, constraints, and policies remain consistent. The process of updating economic analyses allows policies to be altered in step with changes in the economy and in the priorities established for the agricultural sector. Particular objectives can become obsolete or inappropriate as economies grow and change. Low food prices become less important if consumer incomes increase; high producer prices may be unnecessary if farm incomes and production technologies change significantly. Constraints on objectives and policy implementation can alter as well. Developments in the transportation infrastructure, for example, can change the potential for agroindustrial development and for the introduction of new cropping opportunities and can improve the efficacy of producer price support schemes.

In addition to ensuring consistency, quantitative policy analysis can be a dynamic simulation tool to guide patterns of growth and technical change. The development of appropriate technologies has emerged as a growing concern in developing countries. Policy analysis can contribute to discussions on this topic by allowing specification of the changes in relative input requirements necessary for future production technologies. These new technologies reflect combinations of changes in yields through improved seeds and fertilizer, the introduction of new tools or machinery inputs, and changes in the relative use of labor and capital. Discussion with agricultural scientists and engineers can identify which, if any, of the alternatives are technically feasible.

The approach to policy evaluation advanced in this book is built around a simplified analytical framework, the policy analysis matrix

(PAM). The method contains a number of theoretical assumptions and empirical simplifications, and a thorough understanding of its underpinnings is essential for useful application. In most situations, the advantages of the method outweigh its shortcomings. Results are comprehensible to policy-makers and yet are theoretically consistent. The method allows measurement of the effects of policy on producer income as well as identification of transfers among key interest groups—producers in agricultural systems, consumers of food, and policy-makers controlling allocations of the government budget. Results can be easily disaggregated to focus on particular regions, types of farms, or technologies. These items represent critical information for any evaluation of agricultural policy.

The PAM is composed of two sets of identities—one set defining profitabilities and the other defining the difference between private and social values. The selection of an empirical method to estimate PAM is therefore a matter of choice. Traditionally, empirical policy analysis has relied heavily on the estimates of supply and demand curves for various inputs and outputs. In principle, these estimates provide an accurate assessment of market behavior and response. But in practice, sufficient historical data of reliable quality are only rarely available. Even when parameters describing the response to output price changes can be estimated, input demands and the impact of various interventions on production costs are usually overlooked. Further, data are often not sufficiently disaggregated among regions or types of farms. Hence, analysts are unable to assess satisfactorily the impact of government policies on the behavior of a particular commodity system. The resulting analysis is incomplete and often incomprehensible to policy-makers.

This book provides an alternative approach. The methodology is based on the formulation of budgets for representative activities—farming, marketing, and processing—that compose an agricultural commodity system. Private valuations of costs and returns are altered with information about divergences so social costs and returns can be determined. These data are almost always available or can be easily collected, and evaluation can proceed in a timely manner. When reliable information is available for predicting responses of inputs and outputs to social prices, this information can be introduced into the calculation of social costs and returns. But more often, this latter set of adjustments will be made in only an approximate manner. Once these estimations are complete, policy-makers and analysts can decide whether more costly and time-consuming approaches are needed.

Bibliographical Note to Chapter 1

Changes in thinking about development strategies have resulted from the work of a large number of economists. Much of this thinking is summarized in Gerald M. Meier, ed., *Pioneers in Development: Second Series* (New York: Oxford University Press, 1987). Particularly useful essays are those by Gottfried Haberler, "Liberal and Illiberal Development Policy" (pp. 51–83), and Hla Myint, "The Neoclassical Resurgence in Development Economics: Its Strength and Limitations" (pp. 107–36). Two fundamental volumes for changing thought about the particular role of agriculture are Theodore W. Schultz, *Transforming Traditional Agriculture* (1964; reprint, Chicago: University of Chicago Press, 1983); and John W. Mellor, *The Economics of Agricultural Development* (Ithaca: Cornell University Press, 1966). An elaboration of the role of agriculture in the development process is described in John W. Mellor and Bruce F. Johnston, "The World Food Equation: Interrelations among Development, Employment and Food Consumption," *Journal of Economic Literature* 22 (June 1984): 531–74.

One of the first books to focus on the empirical circumstances of agricultural policies was D. Gale Johnson, *World Agriculture in Disarray* (London: Macmillan, 1973). This book is concerned largely with the developed countries. More development-oriented discussions are provided in Theodore W. Schultz, ed., *Distortions of Agricultural Incentives* (Bloomington: Indiana University Press, 1978). Another useful work in this context is C. Peter Timmer, Walter P. Falcon, and Scott R. Pearson, *Food Policy Analysis* (Baltimore: Johns Hopkins University Press, 1983); this book emphasizes the linkages between both macroeconomic and commodity-specific policies and microeconomic responses.

The incorporation of policy distortions into formal economic analysis owes much to the work of Jagdish N. Bhagwati, "The Generalized Theory of Distortions and Welfare," in *Trade, the Balance of Payments, and Growth: Papers in International Economics in Honor of Charles P. Kindleberger,* ed. Bhagwati et al. (Amsterdam: North-Holland, 1971), pp. 69–90; and W. M. Corden, *Trade Policy and Economic Welfare* (Oxford: Clarendon Press, 1974). The empirical framework for policy analysis, based on government objectives, economic constraints, and choice of policy instruments, was first proposed in C. Peter Timmer, "The Political Economy of Rice in Asia: A Methodological Introduction," *Food Research Institute Studies* 14, no. 3 (1975): 191–96. An empirical application of this framework is presented in Scott R. Pearson, J. Dirck Stryker, and Charles P. Humphreys, *Rice in West Africa: Policy and Economics* (Stanford, Calif.: Stanford University Press, 1981), especially pp. 363–95.

The separation of efficiency and nonefficiency objectives remains a hotly debated issue in many areas of economics, particularly in cost-benefit analysis. Most of these arguments have arisen in the context of combining income distribution concerns with efficiency. Chapter 5 of D. W. Pearce, *Cost-Benefit Analysis* (London: Macmillan, 1983), provides a summary of the debate. Analytical models that develop an integrated approach include Robin W. Boadway, "Integrating Equity and Efficiency in Applied Welfare Economics," *Quarterly Journal of Economics* 90 (November 1976): 541–56; and Lyn Squire and Herman G. van der Tak, *Economic Analysis of Projects* (Baltimore: Johns Hopkins University Press, 1975). Prominent opponents of the integration approach include E. J. Mishan, *Introduction to Normative Economics* (New York: Oxford University Press, 1981); and Arnold Harberger, "On the Use of Distributional Weights in Cost-Benefit Analysis," *Journal of Political Economy* 86 (April 1978): S87–S120; and Harberger, "Reflections on Social Project Evaluation," in *Pioneers in Development: Second Series,* ed. Gerald M. Meier (New York: Oxford University Press, 1987), pp. 151–88. Comments on the latter paper by Partha Dasgupta and Deepak Lal, in Meier, *Pioneers in Development,* pp. 189–202, are also directed at the integration issue.

Perhaps the most telling arguments against the incorporation of nonefficiency objectives come from the recent literature on the political economy of public choice. An early contribution to the literature on linking economic benefits and policies with the lobbying efforts of various interest groups is Anne O. Krueger, "The Political Economy of the Rent-Seeking Society," *American Economic Review* 64 (June 1974): 291–303. Additional research on this topic is considered in Robert D. Tollison, "Rent Seeking: A Survey," *Kyklos* 35 (1982): 575–602; and in Jagdish N. Bhagwati, "Directly Unproductive, Profit-Seeking (DUP) Activities," *Journal of Political Economy* 90 (October 1982): 988–1002. Also relevant are the many works on public choice that concern the expression of societal preferences (for example, Dennis Mueller, *Public Choice* [New York: Cambridge University Press, 1979]) and the self-interested actions of government policy-makers (for example, Richard Posner, "Theories of Economic Regulation," *Bell Journal of Economics,* 5 [Autumn 1974]: 335–58).

CHAPTER 2

Introduction to the
Policy Analysis Matrix

THIS CHAPTER explains the construction of the policy analysis matrix and the derivation of measures of efficiency and policy transfer used in agricultural policy analysis. The study of agricultural policy spans three levels—microeconomic behavior of producers, marketing and trade, and macroeconomic linkages. Practitioners of agricultural economics typically give different emphasis to these three topics; micro production issues receive the greatest attention, marketing and trade get less, and macroeconomic links receive little or no coverage. This book argues that excessive specialization precludes successful policy analysis; applied agricultural economists need to understand all of the components of and links among farming systems, domestic and international markets, and macroeconomic policy. Policy analysts have to appreciate feedbacks and tradeoffs within the big picture.

The PAM approach is a system of double-entry bookkeeping. Analysts using PAM have to provide complete and consistent coverage to all policy influences on returns and costs of agricultural production. With this method, applied economists need to be equally capable of analyzing, for example, fertilizer response functions, quantitative restrictions on trade, and real effective exchange rates. The main empirical task is to construct accounting matrices of revenues, costs, and profits. A PAM is constructed for the study of each selected agricultural system—using data on farming, farm-to-processor marketing, processing, and processor-to-wholesaler marketing. The impact of commodity and macroeconomic policies can then be gauged by comparison with the absence of policy.

16

Practical Issues Addressed

Three principal issues—the impact of policy on competitiveness and farm-level profits, the influence of investment policy on economic efficiency and comparative advantage, and the effects of agricultural research policy on changing technologies—can be investigated with the PAM approach. The results can be used to identify what kinds of farmers—categorized by the commodities they grow, the technologies they use, and the agroclimatic zones in which their farms are located— are competitive under current policies affecting crop and input prices and how their profits change as the policies are altered. This issue of farm policy—how agricultural prices affect farming profits—is of primary importance to ministries of agriculture. In the PAM approach, farm budget data (sales revenues and input costs) are collected for the principal agricultural systems. The determination of profit actually received by farmers is a straightforward and important initial result of the analysis. It shows which farmers are currently competitive and how their profits might change if price policies were changed.

A second issue concerns the economic efficiency (or comparative advantage) of agricultural systems and how additional public investment might change the current pattern of efficiency. In what commodity production systems, defined by technology and agroclimatic zone, does the country currently exhibit strong or weak comparative advantage, and how might new investments, using government revenues or foreign aid funds, improve this picture? Investment policy is of primary interest to economic planners who allocate capital budgets, including foreign aid, in attempts to increase efficiency and speed the growth of national income.

With the PAM method, the analyst reassesses the revenues, costs, and profits indicated in farm-level and marketing budgets. Efficiency valuations of outputs and inputs are meant to lead to the highest possible levels of national income. The difference between revenues and costs for a system—both valued in social prices—is social profits, a measure of economic efficiency. New investments that reduce social costs also increase social profits and improve efficiency. An understanding of the array of social profitabilities of agricultural systems greatly reduces the number of detailed benefit-cost analyses needed to evaluate investment alternatives.

A third and closely related set of issues is how best to allocate funds for agricultural research. How can economic analysis be used to help determine the most fruitful directions for primary and applied research to raise crop yields and reduce social costs, thereby increasing social profits? This question is faced by decision-makers in the international agricultural research centers, in several international organizations, and in the agricultural research establishments of certain countries. It is a question also asked by central planners who make allocations to agricultural research budgets.

The approach used in PAM analysis begins with the calculation of existing levels of private (actual market) and social (efficiency) revenues, costs, and profits. This calculation reveals the extent to which actual profits are generated by policy transfers rather than by underlying economic efficiency. Next, agricultural scientists need to project changes in yields and imports resulting from alternative research programs. The effectiveness of such changes can then be gauged by an examination of how they alter private and social profits of current technologies.

The Policy Analysis Matrix

The policy analysis matrix is a product of two accounting identities, one defining *profitability* as the difference between revenues and costs and the other measuring the effects of divergences (distorting policies and market failures) as the difference between observed parameters and parameters that would exist if the divergences were removed. By filling in the elements of the PAM for an agricultural system, an analyst can measure both the extent of transfers occasioned by the set of policies acting on the system and the inherent economic efficiency of the system.

Profits are defined as the difference between total (or per unit) sales revenues and costs of production. This definition generates the first identity of the accounting matrix. In the PAM, profitability is measured horizontally, across the columns of the matrix, as demonstrated in Table 2.1. Profits, shown in the right-hand column, are found by the subtraction of costs, given in the two middle columns, from revenues, indicated in the left-hand column. Each of the column entries is thus a component of the profits identity—revenues less costs equals profits.

Each PAM contains two cost columns, one for tradable inputs and the other for domestic factors. Intermediate inputs—including fertilizer, pesticides, purchased seeds, compound feeds, electricity, transportation, and fuel—are divided into their tradable-input and domestic factor components. This process of disaggregation of intermediate goods or services separates intermediate costs into four categories—tradable inputs, domestic factors, transfers (taxes or subsidies that are set aside

Table 2.1. Policy Analysis Matrix

| | | Costs | | |
	Revenues	Tradable inputs	Domestic factors	Profits
Private prices	A	B	C	D[1]
Social prices	E	F	G	H[2]
Effects of divergences and efficient policy	I[3]	J[4]	K[5]	L[6]

[1]Private profits, D, equal A minus B minus C.
[2]Social profits, H, equal E minus F minus G.
[3]Output transfers, I, equal A minus E.
[4]Input transfers, J, equal B minus F.
[5]Factor transfers, K, equal C minus G.
[6]Net transfers, L, equal D minus H; they also equal I minus J minus K.

Ratio Indicators for Comparison of Unlike Outputs

Private cost ratio (PCR): $C/(A - B)$
Domestic resource cost ratio (DRC): $G/(E - F)$
Nominal protection coefficient (NPC)
  on tradable outputs (NPCO): A/E
  on tradable inputs (NPCI): B/F
Effective protection coefficient (EPC): $(A - B)/(E - F)$
Profitability coefficient (PC): $(A - B - C)/(E - F - G)$ or D/H
Subsidy ratio to producers (SRP): L/E or $(D - H)/E$

in social evaluations), and nontradable inputs (which themselves have to be further disaggregated so that ultimately all component costs are classified as tradable inputs, domestic factors, or transfers).

An example illustrates the process of disaggregating intermediate goods or services. Fertilizer is for most countries a tradable intermediate input. If a particular country is a net importer of fertilizer, the social valuation of a specific kind of fertilizer for its agricultural system is given by the cif (costs, insurance, freight) import price for that fertilizer plus the social costs of moving the input to the representative location in the system. Finding the import price is usually straightforward. Finding the social valuation of the domestic marketing costs is another story, however. It is necessary to study the transportation industry—road or rail—and disaggregate the costs into labor, capital, fuel, and so forth. Each type of cost then needs to be further broken down through use of an appropriate world price and an estimate of local transportation costs.

Private Profitability

The data entered in the first row of Table 2.1 provide a measure of private profitability. The term *private* refers to observed revenues and costs reflecting actual market prices received or paid by farmers, merchants, or processors in the agricultural system. The private, or actual, market prices thus incorporate the underlying economic costs and valuations plus the effects of all policies and market failures. In Table 2.1, private profits, D, are the difference between revenues (A) and costs (B + C); and all four entries in the top row are measured in observed prices. The calculation begins with the construction of separate budgets for farming, marketing, and processing. The components of these budgets are usually entered in PAM as local currency per physical unit, although the analysis can also be carried out using a foreign currency per unit.

The private profitability calculations show the competitiveness of the agricultural system, given current technologies, output values, input costs, and policy transfers. The cost of capital, defined as the pretax return that owners of capital require to maintain their investment in the system, is included in domestic costs (C); hence, profits (D) are excess profits—above-normal returns to operators of the activity. If private profits are negative (D < 0), operators are earning a subnormal rate of return and thus can be expected to exit from this activity unless something changes to increase profits to at least a normal level (D = 0). Alternatively, positive private profits (D > 0) are an indication of supernormal returns and should lead to future expansion of the system, unless the farming area can not be expanded or substitute crops are more privately profitable.

Social Profitability

The second row of the accounting matrix utilizes social prices, as indicated in Table 2.1. These valuations measure comparative advantage or efficiency in the agricultural commodity system. Efficient outcomes are achieved when an economy's resources are used in activities that create the highest levels of output and income. Social profits, H, are an efficiency measure because outputs, E, and inputs, F + G, are valued in prices that reflect scarcity values or social opportunity costs. Social profits, like the private analogue, are the difference between revenues and costs, all measured in social prices—H = (E − F − G).

For outputs (E) and inputs (F) that are traded internationally, the appropriate social valuations are given by world prices—cif import prices for goods or services that are imported or fob export prices for exportables. World prices represent the government's choice to permit consumers and producers to import, export, or produce goods or services domestically; the social value of additional domestic output is thus the foreign exchange saved by reducing imports or earned by expanding exports (for each unit of production, the cif import or fob export price). Because of global output fluctuations or distorting policies abroad, the appropriate world prices might not be those that prevail during the base year chosen for the study. Instead, expected long-run values serve as social valuations for tradable outputs and inputs.

The services provided by domestic factors of production—labor, capital, and land—do not have world prices because the markets for these services are considered to be domestic. The social valuation of each factor service is found by estimation of the net income forgone because the factor is not employed in its best alternative use. This approach requires the commodity systems under analysis to be excluded from social factor price determination. For example, if land is planted to wheat, it cannot grow barley during the identical crop season; the social opportunity cost of the land for the wheat system is thus the net income lost because the land cannot produce barley. Similarly, the labor and capital used to produce wheat cannot simultaneously provide services elsewhere in agriculture or in other sectors of the economy. Their social opportunity costs are measured by the net income given up because alternative activities are deprived of the labor and capital services applied to wheat production.

The practice of social valuation of domestic factors begins with a distinction between mobile and fixed factors of production. Mobile factors, usually capital and labor, are factors that can move from agriculture to other sectors of the economy, such as industry, services, and energy. For mobile factors, prices are determined by aggregate supply and demand forces. Because alternative uses for these factors are available throughout the economy, the social values of capital and labor are determined at a national level, not solely within the agricultural sector. Actual wage rates for labor and rates of return to capital investment are therefore affected by a host of policies, some of which may distort factor prices directly. An enforced and binding minimum-wage law, for example, raises the market wage above what it would have been in the absence of policy and causes observed wages to be higher than the social opportunity cost of labor. But indirect effects can also be important. Distortions of output prices cause different activities to expand or contract, altering in turn the demand and prices of mobile domestic factors.

Fixed, or immobile, factors of production are the factors whose private or social opportunity costs are determined within a particular sector of the economy. The value of agricultural land, for example, is usually determined only by the land's worth in growing alternative crops. Because land is immobile, its value is not directly affected by events in the industrial and service sectors of the economy. But the social opportunity cost of farmland is sometimes difficult to estimate. Within any agroclimatic zone, complete specialization in the most profitable crop is rarely observed. Instead, farmers prefer rotations or intercropping systems that reduce risks of income losses from price variability, yield losses, and pest and disease infestation. Therefore, the social opportunity cost of the land is not accurately approximated by the net profitabilities of a single best alternative crop; instead, it is measured by some weighted average of the social profits accruing from the set of crops planted. Because the correct weights and social profits associated with each crop in the set are generally not known, it is convenient in assessing farming activities to reinterpret crop profits as rents to land and other fixed factors (for example, management and the ability to bear risk) per hectare of land used. This reinterpretation includes private (and social) returns to land as parts of D (and H). Profitability per hectare is then interpreted as the ability of a farming activity to cover its long-run variable costs, in either private or social prices or as a return to fixed factors such as land, management skill, and water resources.

Effects of Divergences

The second identity of the accounting matrix concerns the differences between private and social valuations of revenues, costs, and profits. For each entry in the matrix—measured vertically—any divergence between the observed private (actual market) price and the estimated social (efficiency) price must be explained by the effects of policy or by the existence of market failures. This critical relationship follows directly from the definition of social prices. Social prices correct for the effects of distorting policies—policies that lead to an inefficient use of resources. These policies often are introduced because decision-makers are willing to accept some inefficiencies (and thus lower total income) in order to further nonefficiency objectives, such as the redistribution of income or the improvement of domestic food security. In this circumstance, assessing the tradeoffs between efficiency and nonefficiency objectives becomes a central part of policy analysis.

But not all policies distort the allocation of resources. Some policies are enacted expressly to improve efficiency by correcting for the failure of product or factor markets to operate properly. Market failures occur

Table 2.2. Expanded Policy Analysis Matrix

	Revenues	Costs Tradable inputs	Costs Domestic factors	Profits
Private prices	A	B	C	D[1]
Social prices	E	F	G	H[2]
Effects of divergences and efficient policy	I[3]	J[4]	K[5]	L[6]
Effects of market failures	M	N	O	P
Effects of distorting policy	Q	R	S	T
Effects of efficient policy	U	V	W	X

[1]Private profits, D, equal A minus B minus C.
[2]Social profits, H, equal E minus F minus G.
[3]Output transfers, I, equal A minus E; they also equal M plus Q plus U.
[4]Input transfers, J, equal B minus F; they also equal N plus R plus V.
[5]Factor transfers, K, equal C minus G; they also equal O plus S plus W.
[6]Net transfers, L, equal D minus H; they also equal I minus J minus K; and they equal P plus T plus X.

whenever monopolies or monopsonies (seller or buyer control over market prices), externalities (costs for which the imposer cannot be charged or benefits for which the provider cannot receive compensation), or factor market imperfections (inadequate development of institutions to provide competitive services and full information) prevent a market from creating an efficient allocation of products or factors. Hence, one needs to distinguish distorting policies, which cause losses of potential income, from efficient policies, which offset the effects of market failures and thus create greater income. Because efficient policies correct divergences, they reduce the differences between private and social valuations.

Interpretation of the effects of divergences can be clarified by the expansion of the PAM to include six rows, as shown in Table 2.2. In this expanded PAM, each entry measuring the effects of divergences (I, J, K, and L) is disaggregated into three categories—market failures (fourth row), distorting policies (fifth row), and efficient policies (sixth row). The introduction of efficient policies to offset market failures would change the entries in the first and third rows. To bring about perfect efficiency, a government would introduce efficient policies to offset the effects of market failures and avoid distorting policies, thereby ensuring equality of private and social prices.

In the absence of market failure in the product markets, all divergences between private and social prices of tradable output and inputs are caused by distorting policy. Because the principles are identical for all tradable products, the matrix entries for revenues (tradable outputs) and tradable inputs can be considered together. Output transfers, $I = (A - E)$, and input transfers, $J = (B - F)$, arise from two kinds of policies that cause divergences between observed and world product prices: commodity-specific policies and exchange-rate policy.

Policies that apply to specific commodities include a wide range of taxes or subsidies and trade policy. For example, producer revenues per unit can be raised by producer subsidies (sometimes called deficiency payments in agriculture), tariffs or import quotas on outputs (which raise domestic prices), or domestic price supports enforced by government stockpiling (which require a complementary trade restriction for tradable products). Commodity-specific policies on inputs also affect private profitability. For example, per unit producer costs can be lowered by direct input subsidies or by subsidies on imported inputs.

Typically, PAM accounting is done in domestic currency, but world prices are quoted in foreign currency. Hence, a foreign exchange rate is needed to convert world prices into domestic equivalents. The social exchange rate may differ from observed exchange rates. Undervalued exchange rates reflect an excess supply of foreign exchange that is accumulating as excessive reserves and reducing potential income. Overvalued exchange rates correspond to conditions of excess demand; this demand results in extra foreign borrowing, excessive drawing down of exchange reserves, or rationing of foreign exchange among domestic users.

An overvalued exchange rate is an implicit tax on producers of tradable products because too little domestic currency is earned by exports or paid out for imports. In the absence of commodity policy, the world price of a tradable good determines its domestic price. When the exchange rate is overvalued, the domestic price is lower than its efficiency level and domestic producers are effectively taxed. Undervalued exchange rates exert the opposite effects. Correction for this distortion in PAM is done by conversion of world prices (E and F in the matrix) at the social exchange rate rather than at the official rate. Because exchange rates affect both product prices and factor prices, exchange-rate adjustments are limited to special circumstances—the appearance of multiple exchange-rate regimes or the government's failure to adjust the exchange rate enough to offset the effects of domestic inflation.

The social costs of domestic factors (G) reflect underlying supply and demand conditions in domestic factor markets. Factor prices are thus influenced by the prevailing set of macroeconomic and commodity price policies. In addition, the government can affect factor costs with tax or subsidy policies for one or more of the factors (capital, labor, or land) that create a divergence between private costs (C) and social costs (G). Finally, market imperfections, arising from imperfect information or underdeveloped institutions—which are often characteristic of developing country economies—further influence factor prices. If factor market imperfections exist along with distorting factor policy, both O and S and possibly W are positive components of K. The net transfer, L, thus combines the effects of distorting policy (I, J, and the S part of K) with those of factor market failures (the O part of K) and efficient policies to offset them (the W part of K).

The net transfer caused by policy and market failures (L in the matrix) is the sum of the separate effects from the product and factor markets, $L = (I - J - K)$. (Positive entries in the two cost categories, J and K, represent negative transfers because they reduce private profits, whereas negative entries in J and K represent positive transfers; hence, J and K are subtracted from I, a positive transfer, in the calculation of the net transfer, L.) The net transfer from distorting policy is the sum of all factor, commodity, and exchange-rate policies (apart from efficient policies that offset market failures).

The net transfer can also be found by a comparison of private and social profits. These measures of the net transfer must by definition be identical in the double-entry accounting matrix, $L = (I - J - K) = (D - H)$. Disaggregation of the total net transfer shows whether each distorting policy provides positive or negative transfers to the system. The PAM thus permits comparison of the effects of market failures and distorting policies for the entire set of commodity and macroprice (factor and exchange-rate) policies. This comparison can be made for the complete agricultural system and for each of its outputs and inputs.

Comparisons among Agricultural Systems
Producing Different Outputs

The entries in PAM allow comparisons among agricultural systems that produce identical outputs, either within a single country or across two or more countries. In the accounting matrix, all measures are given as monetary units per physical unit of some commodity. If interest focuses solely on a comparison of one wheat system with another, for example, the matrix entries provide all information necessary for the analysis. Comparisons can be drawn readily by construction of PAM entries for two or more different systems that produce the same quality of wheat. (If necessary, premiums or discounts can be used to correct for quality differences.) Further comparisons can be made between the wheat systems in one country and those in other wheat-producing countries; social exchange rates, incorporating corrections for differential inflation not otherwise offset by exchange-rate changes, are used to convert the other countries' currencies into domestic currency.

Comparisons between wheat and barley—or apples and oranges—are another story, however. To permit comparisons among systems producing different outputs, some common numeraire must be generated. One technique involves the expression of all values relative to a constraining domestic factor resource, such as land. A more common method uses ratios. Both the numerator and the denominator of each ratio are PAM entries defined in domestic currency units per physical unit of the commodity. Therefore, the ratio is a pure number free of any commodity or monetary designation.

Private Profitability

For comparisons of systems producing identical outputs, private profits, $D = (A - B - C)$, indicate competitiveness under existing policies. Construction of a ratio is required to permit comparisons among systems producing different commodities. Direct inspection of the data for private profits is not sufficient. Profitability results are residuals and might have come from systems using very different levels of inputs to produce outputs with widely varying prices. This difficulty might not be apparent in a wheat versus corn example, but it would arise in a comparison of a wheat system with one producing a high-value crop, such as strawberries. This ambiguity is inherent in comparisons of private profits of systems producing different commodities with differing capital intensities.

The problem is circumvented by construction of a private cost ratio (PCR)—the ratio of domestic factor costs (C) to value added in private prices $(A - B)$; that is, $PCR = C/(A - B)$. Value added is the difference between the value of output and the costs of tradable inputs; it shows how much the system can afford to pay domestic factors (including a normal return to capital) and still remain competitive—that is, break even after earning normal profits, where $(A - B - C) = D = 0$. The entrepreneurs in the system prefer to earn excess profits $(D > 0)$, and they can achieve this result if their private factor costs (C) are less than their value added in private prices $(A - B)$. Thus they try to minimize the private cost ratio by holding down factor and tradable input costs in order to maximize excess profits.

Social Profitability

Social profits measure efficiency or comparative advantage. For a comparison of identical outputs, results can be taken directly from the second row of the PAM matrix—social profits equal social revenues less social costs, $H = (E - F - G)$. When social profits are negative, a system cannot survive without assistance from the government. Such systems waste scarce resources by producing at social costs that exceed the costs of importing. The choice is clear for efficiency-minded economic planners: enact new policies or remove existing ones to provide private incentives for systems that generate social profits, subject to nonefficiency objectives.

When systems producing different outputs are compared for relative efficiency, the domestic resource cost ratio (DRC), defined as $G/(E - F)$, serves as a proxy measure for social profits. No new information beyond social revenues and costs is required to calculate a DRC. The DRC plays the same substitute role for social profits as does the PCR for private profits; in both instances, the ratio equals 1 if its analogous profitability measure equals 0. Minimizing the DRC is thus equivalent to maximizing social profits. In cross-commodity comparisons, DRC ratios replace social profit measures as indicators of relative degrees of efficiency.

Policy Transfers

Transfers are shown in the third row of the PAM. If market failures are unimportant, these transfers measure mainly the effects of distorting policy. Efficient systems earn excess profits without any help from the government, and subsidizing policy $(L > 0)$ increases the final level of private profits. Because subsidizing policy permits inefficient systems to survive, the consequent waste of resources needs to be justified in terms of nonefficiency objectives.

Comparisons of the extent of policy transfers between two or more systems with different outputs also require the formation of ratios (for reasons analogous to those offered in the discussions of private and social profits). The nominal protection coefficient (NPC) is a ratio that contrasts the observed (private) commodity price with a comparable world (social) price. This ratio indicates the impact of policy (and of any market failures not corrected by efficient policy) that causes a divergence between the two prices. The NPC on tradable outputs (NPCO), defined as A/E, indicates the degree of output transfer; for example, an

NPC of 1.10 shows that policies are increasing the market price to a level 10 percent higher than the world price. Similarly, the NPC on tradable inputs (NPCI), defined as B/F, shows the degree of tradable-input transfer. An NPC on inputs of 0.80 shows that policies are reducing input costs; the average market prices for these inputs are only 80 percent of world prices.

The effective protection coefficient (EPC), another indicator of incentives, is the ratio of value added in private prices (A − B) to value added in world prices (E − F), or EPC = (A − B)/(E − F). This coefficient measures the degree of policy transfer from product market—output and tradable-input—policies. But, like the NPC, the EPC ignores the transfer effects of factor market policies. Hence, it is not a complete indicator of incentives.

An extension of the EPC to include factor transfers is the profitability coefficient (PC), the ratio of private and social profits or PC = (A − B − C)/(E − F − G), or D/H. The PC measures the incentive effects of all policies and thus serves as a proxy for the net policy transfer, since L = (D − H). Its usefulness is restricted when private or social profits are negative, since the signs of both entries must be known to allow clear interpretation.

A final incentive indicator is the subsidy ratio to producers (SRP), the net policy transfer as a proportion of total social revenues or SRP = L/E = (D − H)/E. The SRP shows the proportion of revenues in world prices that would be required if a single subsidy or tax were substituted for the entire set of commodity and macroeconomic policies. The SRP permits comparisons of the extent to which all policy subsidizes agricultural systems. The SRP measure can also be disaggregated into component transfers to show separately the effects of output, input, and factor policies.

Dynamic Comparative Advantage

The ability of an agricultural system to compete without distorting government policies can be strengthened or eroded by changes in economic conditions. Dynamic comparative advantage refers to shifts in a system's competitiveness that occur over time because of changes in three categories of economic parameters—long-run world prices of tradable outputs and inputs, social opportunity costs of domestic factors of production (labor, capital, and land), and production technologies used in farming or marketing. Together, these three parameters determine social profitability and comparative advantage.

The appropriate world prices for measuring efficiency or comparative advantage are long-run equilibrium levels that approximate best guesses of expected future prices. If the country's decisions to buy or sell on world markets will not have any measurable effect on world price levels, those price levels can be considered exogenous and, once arrived at, can be taken as given for domestic agricultural systems. The world prices are the correct indicators of social valuation of tradable commodities even if a country's decisions to buy or sell internationally do affect the world price of a good. When a large country has market power, however, the analyst needs to take into account the impact of that country's trading decisions on world prices.

In the absence of knowledge of future prices, most analysts project constant long-run real prices rather than fluctuating prices. If new information results in changes in the constant price guess or in the projection of continually increasing or decreasing future prices, these changes can be incorporated easily into the PAM. Separate PAMs can be constructed for each year, and each can have different assumed world prices.

Costs of factor services in any country can be expected to change over time. But cyclical variations in the real wage and the real return to capital, associated with swings in macroeconomic policy, are not the primary focus of the PAM method. Instead, interest centers on long-run trends in the costs of labor, capital, and land. As economies grow, real wages typically rise, both in absolute terms and relative to real costs of capital and land. For agricultural systems, changes in the social opportunity costs of labor and of capital depend on changes in the national environment for investment and growth. Land rental rates are endogenous to agriculture but will be constrained by changes in world prices and in real wage and interest rates, because payments to land and other permanently fixed factors come out of profits. Analysis of projected comparative advantage therefore includes both the future pressures that changing real factor prices might exert on agricultural systems and the influences of likely world prices for tradable outputs and inputs. The results identify systems that can readily expand and those that will have to contract or change in order to survive.

Changes over time in factor and commodity prices can also influence agricultural technologies. Farmers and researchers innovate, often by finding new ways of using less of factors that are relatively expensive (usually labor) and more of other inputs. Successful technological change permits commodities to be produced with reduced costs of one or more inputs. Empirical analysis of intrasystem change can be done with partial budgeting, a technique in which individual cost-saving or revenue-increasing changes can be analyzed within the PAM for the initial system.

Concluding Comments

The central purpose of PAM analysis is to measure the impact of government policy on the private profitability of agricultural systems and on the efficiency of resource use. Private profitability and competitiveness are likely to be uppermost in the minds of those concerned specifically with agricultural incomes. Social profitability and efficiency are often emphasized by economic planners whose concern is the allocation of resources among sectors and the growth of aggregate income in the economy. Both sets of issues ultimately focus on the incentive effects of policy—part of the difference between private and social profitability—and on how policy incentives might be altered. Through evaluation of private and social revenues and costs, the PAM method is designed to illuminate these related issues of agricultural policy analysis. The approach is particularly well suited to empirical analysis of agricultural price policy and farm incomes, public investment policy and efficiency, and agricultural research policy and technological change.

The PAM approach to policy evaluation advocates a disaggregated view of efficiency effects (as measured by social profitability) and of nonefficiency effects. The analyst can do much in describing the contributions of a particular system to nonefficiency objectives and in quantifying implications for efficiency (aggregate income gains or losses). But it is left to the discretion of each policy-maker to determine whether tradeoffs between efficiency and nonefficiency objectives merit changes in policy or maintenance of incentives to particular systems.

In other approaches to policy analysis, it is desirable to aggregate measures of efficiency and nonefficiency effects into a single measure. Income distribution concerns, for example, can be introduced into the social cost estimates by weighting (with a value less than 1) of the efficiency-determined value of unskilled labor wages. Concerns for food self-sufficiency can be introduced by the addition of a premium to the world market value of output. If these weights were incorporated in the calculations of social profitability, the policy-makers' decision would become automatic and predetermined: encourage all systems with positive social profitability and discourage all systems with negative profitability.

The disadvantage of the aggregate approach lies in its tendency to lump together high-quality information (observable data on prices and input-output relationships) with relatively poor-quality information (implicit weights of society or of policy-makers regarding various prices of inputs and outputs). Moreover, attempts to quantify implicit policy weights presume the existence of some dictatorial policy-maker who speaks on behalf of society. Policy-making rarely occurs in such an environment. Policies are the outcomes of negotiated conflict between interest groups both within and outside the government. Quantitative studies provide improved information and thus increase the probability of good policy decisions. But these decisions, and the tradeoffs implied between efficiency and nonefficiency objectives, are the outcomes of debate based on this information, not inputs into the collection of information.

Bibliographical Note to Chapter 2

The policy analysis matrix approach to agricultural policy analysis was originally developed in 1981 by the authors and several colleagues to establish a framework for studying changes in agricultural policies in Portugal. See Scott R. Pearson et al., *Portuguese Agriculture in Transition* (Ithaca: Cornell University Press, 1987).

The PAM approach, like all methodologies, has many antecedents. Its links to the extensive literature on social benefit-cost analysis (SBCA) were made apparent in this chapter. Applications of these analytical techniques to agricultural investment projects are discussed in J. Price Gittinger, *Economic Analysis of Agricultural Projects,* 2d ed. (Baltimore: Johns Hopkins University Press, 1982), which also contains an extensive bibliography of the theoretical literature on SBCA. Because the PAM is a way of carrying out both efficiency and policy analysis, the approach also has antecedents in the literature on international trade policy. That literature is summarized, with full listings of citations, in two chapters of Ronald W. Jones and Peter B. Kenen, eds., *Handbook of International Economics* (Amsterdam: North-Holland, 1984): Anne O. Krueger, "Trade Policies in Developing Countries," pp. 519–69; and Robert E. Baldwin, "Trade Policies in Developed Countries," pp. 571–619.

A methodological breakthrough underpinning the eventual development of the PAM was the invention of the domestic resource costs (DRC) approach for measuring social profitability. The DRC was developed independently by Michael Bruno in Israel and Anne Krueger in the United States; early publications demonstrating its use are Michael Bruno, "The Optimal Selection of Export-Promoting and Import-Substituting Projects," in Bruno, *Planning the External Sector: Techniques, Problems, and Policies* (New York: United Nations, 1967);

and Anne O. Krueger, "Some Economic Costs of Exchange Control: The Turkish Case," *Journal of Political Economy* 74 (October 1966): 466–80. The early DRC literature is reviewed in Scott R. Pearson, "Net Social Profitability, Domestic Resource Costs, and Effective Rate of Protection," *Journal of Development Studies* 12 (July 1976): 320–33.

During the 1970s and the first half of the 1980s, a series of detailed empirical studies of comparative advantage in agriculture, based in part on the DRC method, was carried out by faculty members and graduate students at the Food Research Institute, Stanford University, and by their colleagues from other institutions. Some of the results of these investigations are reported in Scott R. Pearson and John Cownie, *Commodity Exports and African Economic Development* (Lexington, Mass.: D. C. Heath, 1974); Eric Monke, Scott R. Pearson, and Narongchai Akransee, "Comparative Advantage, Government Policies, and International Trade in Rice," *Food Research Institute Studies* 15 (1976): 257–83; Scott R. Pearson et al., *Rice in West Africa: Policy and Economics* (Stanford, Calif.: Stanford University Press, 1981); Walter P. Falcon et al., *The Cassava Economy of Java* (Stanford, Calif.: Stanford University Press, 1984); C. Peter Timmer, ed., *The Corn Economy of Indonesia* (Ithaca: Cornell University Press, 1987); and Pearson et al., *Portuguese Agriculture*. Each volume contains discussions of methods and applications as well as citations of relevant studies.

The principal motivation for combining the DRC approach with measures of policy transfers is drawn from W. M. Corden, *Trade Policy and Economic Welfare* (Oxford: Clarendon Press, 1974). The distinction between efficient and distorting policy and the definition of divergences as including the effects of distorting policies and uncorrected market failures are two central insights in that book. The empirical approximation of the effects of policy being the difference between actual market (private) and efficiency (social) valuations was first made in the study of rice in West Africa, Pearson et al., *Rice in West Africa*. The closest forerunner to the complete PAM approach is the method used in William D. Ingram and Scott R. Pearson, "The Impact of Investment Concessions on the Profitability of Selected Firms in Ghana," *Economic Development and Cultural Change* 29 (July 1981): 831–39.

A standard reference on the theory of effective protection is W. M. Corden, *The Theory of Protection* (Oxford: Oxford University Press, 1971). A useful updating of that work, complete with bibliographic listings, is W. M. Corden, "Effective Protection Revisited," in Corden, *Protection, Growth and Trade: Essays in International Economics* (Oxford: Basil Blackwell, 1985), pp. 141–53.

Dynamic comparative advantage is discussed in two early papers: Hollis B. Chenery, "Comparative Advantage and Development Policy," *American Economic Review* 51 (March 1961): 18–51; and Michael Bruno, "Development Policy and Dynamic Comparative Advantage," in *The Technology Factor in International Trade*, ed. Raymond Vernon (New York: Columbia University

Press, 1970), pp. 27–64. An effort to measure dynamic comparative advantage in agricultural systems is reported by Roger Fox and Timothy J. Finan, "Patterns of Technical Change in the Northwest" and "Future Technical and Structural Adjustments in Northwestern Agriculture," in Pearson et al., *Portuguese Agriculture*, pp. 187–220.

PART II

EVALUATION OF
AGRICULTURAL POLICY

CHAPTER 3

Commodity Policy

AGRICULTURAL POLICY is most commonly associated with the set of commodity-specific actions that cause domestic prices of agricultural products to differ from their counterpart world prices. Recognizing that farmers respond to profits as well as to output prices, most governments also use instruments to influence the costs of purchased inputs, such as fertilizer and fuel. A central message of the PAM approach is that policy-makers could make more effective policies if they directly considered macroeconomic prices—exchange rates and factor prices (interest, wages, and land rental rates)—in their agricultural decisions. This chapter focuses on the instruments of mainstream agricultural policy—the setting of prices for farm products and commodity inputs.

Governments use an array of commodity-specific instruments to influence product prices. The tangible economic objectives for the agricultural sector of most governments, especially those of developing countries, are to promote economic efficiency (and hence higher incomes), to distribute incomes, to provide food price stability and security of food supplies, to create conditions of adequate nutritional status for all, and to contribute to fiscal balance in the public sector. The effects of commodity policies can thus be analyzed by measurement of their influences on each of these objectives.

Classification of Price Policies

The influences of commodity price policy can be studied under several simplifying assumptions. Because emphasis is on national government actions, it is convenient to illustrate the effects of different commodity policies with diagrams depicting the price and quantity of a single commodity in a national market. It is assumed that the commodity is internationally tradable. Whether the country is an importer or an exporter and how much it imports or exports are issues addressed in the analysis, but the assumption is that international transfer costs are small enough to allow profitable foreign trade. Indeed, for simplicity of presentation, these costs are assumed to be zero so that the world price represents both the cif price for imports and the fob price for exports. Domestic marketing margins, market failures, and changes in demand and supply of related commodities are also ignored, again for ease of exposition. An additional assumption affecting the world price is that the country is sufficiently small in the world market that its international sales or purchases do not measurably affect the world price. This assumption about the smallness of the country is usually not crucial, but it permits simple graphical illustrations. In this circumstance, the country can import or export any quantity of the commodity at a constant world price, which can be represented by a horizontal line.

A classification of commodity price policies is presented in Table 3.1 to help explain the effects of policy changes. The table lists ten types of price policies, distinguished according to three criteria—type of instrument (subsidy or trade policies), intended beneficiary (producers or consumers), and type of commodity (importable or exportable).

Type of Instrument

The first criterion is the distinction between subsidy policies and trade policies. A subsidy is a payment from the government treasury. (A tax, which is a payment into the treasury, is thus a negative subsidy.) The rate of subsidy is the amount paid per unit of subsidized output, and the total subsidy is calculated by multiplication of that rate by the amount of subsidized production or consumption. The purpose and effect of a subsidy is to create domestic prices that differ from world prices; sometimes policies create separate domestic prices for producers and consumers, beyond the difference caused by a marketing margin.

A trade policy is a restriction placed on imports or exports of a commodity. The restriction can be applied to either the price of a tradable commodity (with a trade tax) or its quantity (with a trade quota) to reduce the amount traded internationally and to drive a wedge between the world price and the domestic price. For imports, trade policy imposes either a per unit tariff (import tax) or a quantitative

Table 3.1. Classification of Commodity Price Policies

Instrument	Policies affecting producers	Policies affecting consumers
Subsidy policies:	Producer subsidies	Consumer subsidies
a. That do not change domestic market prices	On importables (S+PI, S−PI)	On exportables (S+CE, S−CE)
b. That do change domestic market prices	On exportables (S+PE, S−PE)	On importables (S+CI, S−CI)
Trade policies (all of which change domestic market prices)	Restrictions on imports (TPI)	Restrictions on exports (TCE)

S = Subsidy policy (+ = positive subsidy, − = tax).
T = Trade policy.
P = Affecting producers.
C = Affecting consumers.
I = Of importables.
E = Of exportables.

restriction (import quota) to limit the quantity imported and raise the domestic price above the world price. Trade policy for exports involves a limitation on the quantity exported through imposition of either a per unit export tax or an export quota, causing the domestic price to be lower than the world price.

Trade policies differ from subsidy policies in three respects. The first involves implications for the government budget. Trade restrictions either benefit or have no impact on the budget, whereas subsidies always use the government treasury to drive a wedge between world and domestic prices. With negative subsidies, the treasury benefits; with positive subsidies, the treasury loses. Some subsidies alter prices for domestic producers while permitting consumers to continue to purchase at the world price; other subsidy policies permit producers to continue to receive the world price but alter prices for consumers. Some subsidy policies affect both groups. As a result, the effects of subsidies on the treasury vary widely in magnitude.

The second difference is reflected in the number of alternative subsidy and trade policies. Eight types of subsidy policies exist because either producer or consumer subsidies can be applied to both importables and exportables. In each instance, subsidies can be positive or negative. However, there are only two basic types of trade policies—restrictions on imports and controls on exports. Trade flows of imports or exports can be restricted by trade taxes or quota policies so long as a government has effective mechanisms to control smuggling. The opposite effect, expansion of imports or exports, cannot, however, be created by trade regulations. Countries can and do subsidize imports or exports and expand their trade movements. But either of these actions is actually a subsidy policy and is already classified as such.

The final difference between subsidy and trade policies concerns the extent of their applicability. All goods and services are either tradable or nontradable, depending on comparative levels of domestic costs of production and international transfer costs. By definition, a trade policy can apply only to a tradable good, since restrictions can be imposed only if trade flows exist. Subsidy policies, however, can be applied to all goods, including nontradables. Initially, the discussion of policy effects is restricted to tradables—importables and exportables—to permit comparison of subsidy and trade policies.

Beneficiary Group

The second criterion for policy classification is whether the policy is intended for producers or consumers. A subsidy or trade policy causes transfers among producers, consumers, and the government treasury. Unless the government budget pays for the entire transfer, when producers gain, consumers lose, and when consumers gain, producers lose. It is tempting to describe this situation as a zero-sum game in which gains for one group are just offset by losses for another. But the transfers are accompanied by efficiency losses, so gainers gain less than losers lose. Hence, the benefits for one group (producers, consumers, or the government treasury) are less than the sum of the losses for the other groups.

Type of Commodity

The third classifying device is a distinction between importables and exportables. When there is no price policy, the price prevailing domestically is the world price—the cif import price for an importable or the fob export price for an exportable. Hence, a country imports some of its importables and exports some of its exportables unless the government intervenes. Producers gain and consumers lose with policies that raise domestic prices for either importables or exportables, whereas consumers benefit and producers are hurt with policies that lower domestic prices of either kind of tradable good.

In Table 3.1, the three classification criteria are described with a shorthand notation, given in parentheses with each entry in the table. The first letter in each three-letter symbol refers to whether the entry is a subsidy policy, S, or a trade policy, T; a superscript + is added to the S to denote positive subsidies, and a superscript − is used to denote negative subsidies (taxes). The second letter signifies whether the policy is intended to affect producers, P, or consumers, C; and the third letter identifies the affected commodity as being an importable, I, or an exportable, E. For example, the policy to restrict imports is labeled TPI, because it is a trade policy benefiting producers by raising the domestic price of importables.

Illustrations of Distorting Price Policies

Figure 3.1 contains representations of the four types of positive subsidy policies. The four types of negative subsidy policies are simply the converse of the positive ones and thus are not illustrated. Figure 3.2 illustrates the two kinds of trade policies. Any price policy for tradable commodities can be identified as one of these ten basic types, allowing prediction of the kinds and directions of policy effects. Three effects of price changes are of principal interest in agricultural policy analysis: the quantities of the commodity that are produced, consumed, and traded (imported or exported); the income transfers to or from producers, consumers, and the government budget; and the efficiency losses in production or consumption. The likely magnitudes of change depend on the size of the price change and on the elasticities of demand and supply, which compare percentage changes in quantities with percentage changes in prices. Because of data limitations, reliable estimates of elasticities are often difficult to obtain. But the direction of change is predictable so long as the demand curve is downward sloping and the supply curve is upward sloping.

Policies for Importables

A positive producer subsidy on importables, S^+PI, is illustrated in Figure 3.1a. A government might desire to expand domestic output of an agricultural crop and choose to subsidize its production from budgetary revenues. The policy raises the domestic payment to producers, P_P, to a level above the world price, P_W, thereby increasing the quantity of local output from Q_1 to Q_2. Local consumption (Q_3) is not affected. Under an S^+PI policy, the domestic market price, P_D, remains equal to

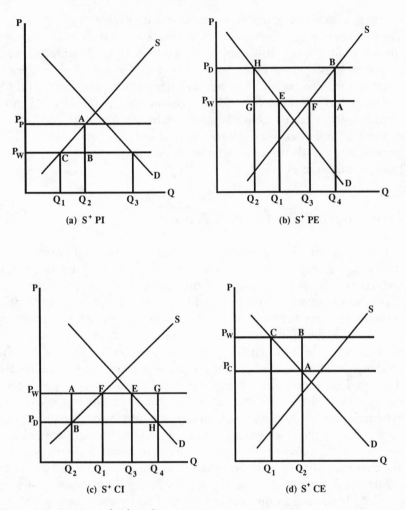

Figure 3.1. Positive subsidy policy

the world price, P_W. The per unit revenue received by producers, P_P, is higher than P_D by the amount of subsidy from the government. The subsidy scheme is feasible only if producers and consumers are separated by enough economic distance, product transformation, or administrative control that the commodity cannot be repurchased at the lower market price and resold at the higher producer price. The quantity of imports is reduced from $Q_3 - Q_1$, the level without policy, to $Q_3 - Q_2$.

The level of subsidy per unit, $P_P - P_W$, is applied to Q_2 of production, so the total transfer to producers from the government budget is Q_2 x $(P_P - P_W)$, shown by the area P_PABP_W. This budget transfer creates an

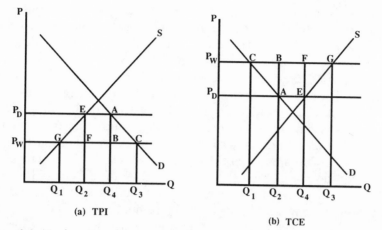

(a) TPI

(b) TCE

Figure 3.2. Trade policy

efficiency loss in the economy, because the government chooses not to permit scarce resources to be allocated by a price formed at P_W, the competitive equilibrium. The government has the option of obtaining $Q_2 - Q_1$ by importing; the opportunity cost of this amount of output is given by its import cost, Q_1CBQ_2. The subsidy policy causes additional local production to replace imports; the value of the local resources used to produce the $Q_2 - Q_1$ of output is represented by the area under the supply curve, Q_1CAQ_2. The efficiency loss from policy S^+PI is the difference between the resource cost of increased domestic production, Q_1CAQ_2, and the opportunity cost, Q_1CBQ_2, or the triangle CAB.

A positive consumer subsidy on importables, S^+CI, is portrayed in Figure 3.1c. It differs significantly from S^+PI in that it results in a single domestic price for both producers and consumers, P_D, at a level lower than the world price, P_W. The policy provides a per unit subsidy of $P_W - P_D$ on imports. Production is reduced from Q_1 to Q_2, consumption increases from Q_3 to Q_4, and imports are raised from $Q_3 - Q_1$ to $Q_4 - Q_2$. The transfers from S^+CI consist of two parts. The subsidy, $(P_W - P_D)(Q_4 - Q_2)$, or AGHB, is transferred from the government budget. Because the subsidized imports reduce the price for producers as well as consumers, producers transfer P_WABP_D to consumers.

Efficiency-related losses are now present in both production and consumption. In production, the reduction of output from Q_2 to Q_1 implies a loss of income of $P_W(Q_2 - Q_1)$, or Q_2AFQ_1. By reducing output, the economy saves inputs; the value of these resources is again represented by the area under the supply curve, Q_2BFQ_1. The net income loss to the economy is represented by AFB.

Like its counterpart in production, the efficiency loss in consumption results from the government's decision to set the consumer price below the world price. The justification for this consumption loss is different from that for production. By subsidizing imports and lowering the consumer price, the government causes consumption to increase from Q_3 to Q_4. The opportunity cost (or value) of this increment in consumption is measured at world prices, $P_W(Q_4 - Q_3)$, or area Q_3EGQ_4. Total willingness to pay for this increment in consumption is approximated by the area under the demand curve, just as total costs of production are represented by the area under the supply curve. The consumer efficiency loss in consumption of policy S^+CI is therefore the difference between the opportunity cost of the increased consumption, Q_3EGQ_4, and the willingness to pay for the increased consumption, Q_3EHQ_4, or the triangular area EGH. Because the consumer efficiency loss is defined in terms of willingness to pay, it does not represent forgone income.

A restriction on imports, TPI, is shown diagrammatically in Figure 3.2a. This policy benefits producers by raising the domestic price facing both producers and consumers, P_D, above the world price, P_W, thereby allowing domestic output to expand from Q_1 to Q_2. The trade effect is a reduction in imports from $Q_3 - Q_1$ to $Q_4 - Q_2$, reflecting the increase in local supply and the decrease in local demand. The policy is implemented by imposition of a per unit tariff of $P_D - P_W$ or an equivalent quantitative restriction permitting imports at a level of $Q_4 - Q_2$, which would have the same restrictive effect on trade as a tariff of $P_D - P_W$. Unlike S^+PI, which does not affect consumers, TPI causes a reduction in consumer demand from Q_3 to Q_4. Because of higher prices, consumers transfer income of $(P_D - P_W)Q_4$, or P_DABP_W, to producers $(P_D - P_W)Q_2$, or P_DEFP_W, and to the government budget $(P_D - P_W)(Q_4 - Q_2)$, or FEAB. Consumer efficiency losses are again measured as the difference between the opportunity cost of the change in consumption, $P_W(Q_3 - Q_4)$, or Q_4BCQ_3, and the willingness to pay for the same increment, Q_4ACQ_3. The consumer efficiency loss is thus the triangular area ABC. Efficiency losses in production (EFG) are measured in the same way as in the S^+PI case.

Policies for Exportables

The three types of policies just discussed influence importables. Similar detailed discussion for exportables is not necessary, because the three types of policies that affect exportables are obverses of the policies for importables. In Figure 3.1, S^+PI is the opposite of S^+CE, and S^+CI is matched with S^+PE; in Figure 3.2, TPI and TCE are opposites. All that

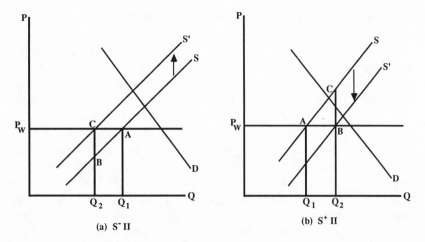

Figure 3.3. Input policy

needs to be done tu apply the discussion on policies for importables to policies for exportables is to exchange *exportables* for *importables, decrease* for *increase, more* for *less,* and so forth. Indeed, one can see the diagram for either member of the pair by taking a thin piece of paper on which is drawn the diagram for its partner policy, turning the paper over so that the original bottom is at the top, and looking into the light through the back side of the paper. The measures of efficiency losses from exportable policies are labeled accordingly. This exercise demonstrates the totally symmetric relationship between each pair of policies.

Policies for Tradable Inputs

Tradable-input policies are identified in the classification system with the addition of I (inputs) to the second criterion. Tradable-input policies have direct relevance only for producers of outputs; therefore, a superscript + on the first criterion describes input policies that encourage input use and benefit producers; a superscript − describes input policies that discourage input use and are costly to producers. For example, a fertilizer subsidy for rice producers would be classified as S^+II or S^+IE, because the input policy has reduced the cost of producing a given quantity of rice. A negative fertilizer subsidy (tax) raises production costs and is designated S^-II or S^-IE. Trade policies bear the designations T^+I_- and T^-I_-.

The efficiency effects of input policies on a particular output market are illustrated in Figure 3.3. Figure 3.3a shows the effect of a tax on an input used to produce an importable output. For each level of output, the costs of production are increased. The magnitude of increase depends on the share of the input in production costs and the ability of the producer to substitute other inputs for the taxed input. A tax on ammonium phosphate, for example, will have little impact on the costs of producing rice if rice producers are able to substitute alternative fertilizers, such as urea and superphosphate. A tax on all fertilizers will have a larger impact, because the rice producer cannot avoid the tax. The only possible response is to reduce fertilizer use and decrease output.

In Figure 3.3a, output declines from Q_1 to Q_2, the intercept of the new output supply curve, S', with the world price. The efficiency loss to the economy is measured by the triangular area ABC; this area represents the difference between the value of lost output, Q_2CAQ_1, and the cost of producing that output, Q_2BAQ_1. If the input is used in more than one output market, similar efficiency losses result in those markets. By summing across all the output markets, one can estimate the total efficiency effects of the input tax.

Figure 3.3b shows the impact of an input subsidy on the production of an importable output. Costs of producing any given amount of output are reduced, and the lower input price encourages intensified use. The supply curve in the output market shifts downward; production increases from Q_1 to Q_2. The efficiency cost of this policy is ABC— the difference between the cost of producing the increased output, Q_1ACQ_2, and the value of the increased output, Q_1ABQ_2. As before, total efficiency effects can be estimated by consideration of all outputs that use the subsidized input.

Policies for Nontradables

Nontradable-commodity policies (N) are the final class of commodity price policies. By definition, trade policies cannot affect nontradables; therefore, all nontradable policies are designated subsidy (S) policies. But unlike tradable-commodity policies, nontradable subsidies affect both producers and consumers. All nontradable production is consumed domestically (assuming no permanent storage or destruction of the commodity). Therefore, any policy that encourages production (such as a positive subsidy to producers) also results in lower consumer prices; these lower prices are necessary to clear the market. Similarly, policies that discourage production cause higher consumer prices. Because both groups either benefit or incur efficiency losses, the second criterion is unnecessary. Instead, the superscript + or − on the first criterion is sufficient to indicate whether the subsidy policy is encouraging production and consumption (S^+) or discouraging them (S^-).

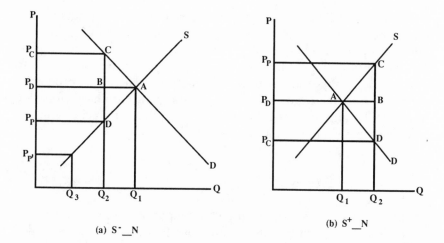

Figure 3.4. Nontradable policy

Figure 3.4 illustrates the effects of nontradable-commodity policies on consumer and producer efficiency. In Figure 3.4a, a tax (equal to $P_C - P_P$) on the commodity is introduced. Whether the tax is initially imposed on consumers or producers makes no difference. In the new equilibrium, the two groups share the burden of the tax because only one level of output, Q_2, is consistent with a per unit tax of $P_C - P_P$. Attempts to force the entire amount of the tax on producers ($P_D - P_{P'}$), for example, would cause output to decline below Q_2 to a point such as Q_3. At this point, consumer prices would be well above P_C and the producer price after taxes would be above $P_{P'}$. Subsequently, output would increase and consumer prices would fall until the amount of the tax accounted for the full difference between consumer and producer prices.

Relative to the pretax situation, consumer prices have increased (to P_C) and producer prices have fallen (to P_D). Efficiency losses are measured by a comparison of consumers' willingness to pay with production resource costs to the value of forgone output, Q_2BAQ_1. Consumer efficiency losses are measured as BCA and production efficiency losses as DBA.

Analysis of positive subsidies for the nontradable commodity, S^+_N, is presented in Figure 3.4b. Both groups necessarily benefit from the subsidy. Output expands from Q_1 to Q_2. Relative to initial price P_D, the consumer price is lower and the producer price is higher. Efficiency losses are measured by comparison of the value of increased output at initial prices, Q_1ABQ_2, to the incremental production cost and the increment in consumer willingness to pay. The total efficiency loss is ACD.

Analysis of the effects of input subsidy policies on nontradable outputs proceeds in a fashion similar to that described in Figure 3.3. Positive input subsidies, S^+IN, cause the nontradable-output supply curve to shift downward; input taxes, S^-IN, cause the curve to shift upward. But, unlike the case of tradable outputs (where only producers are affected by input policies), both consumer and producer prices of nontradable outputs change. Hence, the measures of efficiency losses assume shapes similar to those shown in Figure 3.4.

Multiple Interventions

The analysis of commodity output and input policies has focused on the effects of single interventions. But governments can, and often do, impose multiple interventions on commodity systems. In these cases, the effects of the component commodity policies are added together to identify gainers and losers from policy. With multiple interventions, consumers and producers may both gain or both lose.

Figure 3.5 illustrates the possibility for tradable-commodity subsidy policies. In Figure 3.5a, both consumers and producers receive positive subsidies; this panel is a combination of the two policies described in Figures 3.1a and 3.1c. The import subsidy policy increases consumption from Q_4 to Q_3 and, in isolation, would reduce domestic production from Q_1 to Q_2'. But with an S^+PI policy introduced for producers, production increases from Q_1 to Q_2'. Total imports, $Q_3 - Q_2'$, could be greater or smaller than imports before policy. Transfers from the treasury go to consumers, P_WABP_C, and to producers, P_PEFP_W. The government pays two subsidies for domestic production, one to producers and one to consumers. Efficiency losses are found in consumption (GAB) and in production (HEF).

In Figure 3.5b, both groups are taxed relative to the world price. Governments impose a tax on consumption of the commodity (such as a trade tax), raising consumer prices from P_W to P_C. Total consumption declines from Q_4 to Q_3. Conducted in isolation, this policy (TPI) would benefit domestic producers. But the government is assumed to be capable of imposing a producer price, P_P, that is below the world price (S^-PI), causing domestic production to decline from Q_1 to Q_2. As before, the effect of the dual policies could either raise or lower trade volume. But in this case, the government treasury gains—FEBC on imports, P_WP_CEF on consumption of domestic production, and the tax on production, P_PP_WFG. Domestic production is taxed twice, with the

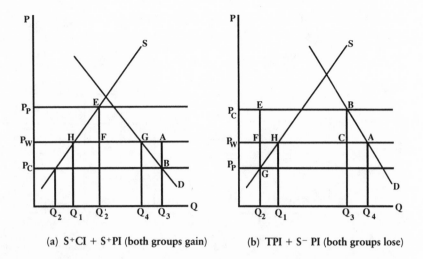

(a) S⁺CI + S⁺PI (both groups gain)   (b) TPI + S⁻ PI (both groups lose)

Figure 3.5. Multiple interventions—commodity subsidy policies

government receiving one part from domestic producers and the other from domestic consumers. Efficiency losses are ABC (consumers) and FGH (producers).

In this example, multiple interventions maintain or increase the sources of efficiency losses; both producers and consumers incur such losses. But it is possible for subsequent interventions to exercise a more benign role, offsetting part of the efficiency losses created by the initial policy intervention. Figure 3.6 illustrates this result. Figure 3.6a reproduces the S⁺CI policy of Figure 3.1c. Initially, both taxpayers and producers suffer efficiency losses as a consequence of import subsidies. But in Figure 3.6a, the government is assumed to implement a price support program in which producers are offered the world price. In this case, production remains at Q_1, the output level dictated by world prices, and the economy avoids the production efficiency loss (BAF). Only the taxpayer efficiency loss (EGH) remains. Because the producer price program does not subsidize producers relative to world prices, it is classified as a benign transfer.

Figure 3.6b reproduces the S⁺PE policy of Figure 3.1b, the remaining case in which a positive subsidy policy creates efficiency losses in production and consumption. In this case, the benign transfer policy is a subsidy to domestic consumption, so that consumers pay world prices and only producers receive the higher domestic price. The combination of policies avoids the consumer efficiency loss (GHE), leaving the economy with only the producer efficiency loss (FBA).

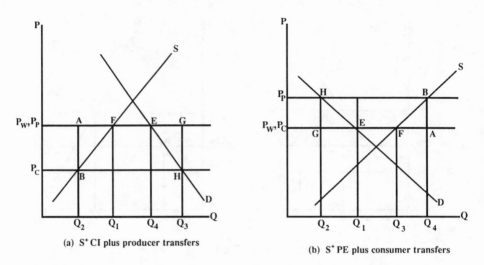

(a) S⁺ CI plus producer transfers

(b) S⁺ PE plus consumer transfers

Figure 3.6. Multiple interventions: Commodity policy plus benign transfer policy

PAM and the Evaluation of Commodity Policy

The PAM approach to the measurement of efficiency gains or losses differs from that used in the partial equilibrium diagrams. The PAM measure of efficiency is social profit—the net change in national income that results from the introduction of the commodity system into the economy. The partial equilibrium diagrams, on the other hand, measure efficiency effects as the difference between incremental social benefits and incremental social costs. However, the diagrams are more useful as heuristic tools, and both approaches yield similar characterizations of policy effects.

Figure 3.7 compares the partial equilibrium measures with those contained in the PAM. The marginal cost curve of the representative firm is shown as UVWXYS. For convenience, only one policy is assumed present—a tax on production of an exportable (S−PE). This policy causes the domestic producer price (P_P) to be less than the world price (P_W). Total output is therefore Q_P instead of Q_W. The partial equilibrium measure of the efficiency cost of this policy is measured as the triangular area above $Q_W YXQ_P$.

The PAM measures of revenues, costs, and profits are presented below the diagram. The measures of private and social revenues are the rectangular areas bounded by the relevant prices and quantities. The measures of total costs of production are the areas under the marginal cost curve: OUVWXQ_P for the distorted domestic price, and OUVWX-

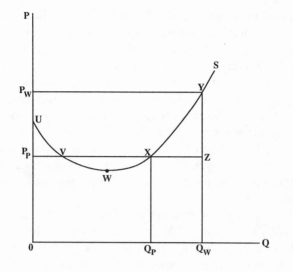

Policy Analysis Matrix

	Revenues	Costs	Profits
Private	$OP_P XQ_P$	$OUVWXQ_P$	$VWX-UVP_P$
Social	$OR_W YQ_W$	$OUVWXYQ_W$	$P_W UVWXY$
Effects of Divergences	$-P_W YZP_P$	$-Q_W ZXQ_P$	$-P_W YXP_P$
	$-Q_W ZXQ_P$	$-XYZ$	

Figure 3.7. Comparisons of PAM and partial equilibrium measures of policy effects

YQ_W for the world price. All the remaining PAM entries are obtained from these four values: revenues minus costs yields profits; private minus social values represents the impact of the divergence on revenues, costs, and profits. Relevant areas are denoted in the figure.

The difference between PAM and the partial equilibrium measures of efficiency arises because social profit captures the total contribution of the system to national income, whereas the partial equilibrium measure is concerned with the incremental impact of the price policy. In the PAM the incremental impact of policy (the triangular area above $Q_W YXQ_P$) enters as part of L, the difference between private and social profit (or, equivalently, the net effect of policy on the system). The value of L thus includes the net transfers from producers as well as efficiency losses. In Figure 3.7, L is measured as the (negative) area, $-P_W YXP_P$.

Separation of the transfer and the incremental efficiency effects is done easily with additional information about consumption and trade under observed (distorted) market conditions. Information on domestic consumption, trade volume, and the type of policy instrument allows classification of the producer transfers among consumers and the government budget. In Figure 3.7, for example, producer taxes could be attained by a general sales tax, providing no benefits for consumers and causing the government to capture all of the transfer. The producer price could also be attained by an export tax; in this case, the government garners revenue only from the quantity exported; the remainder of the transfer goes to domestic consumers because domestic market prices are below world prices.

The government budget transfer is often critical to the choice of a particular policy instrument. The desire to provide public services and control deficits means that governments face budget constraints; in turn, policy decisions often are taken principally, or even wholly, on the basis of their budgetary implications. A common explanation for the widespread use of trade restrictions instead of subsidies is the budgetary difference between the two types of policies. Trade taxes raise revenue (unless they eliminate trade) and force consumers to provide income transfers both to producers and to the treasury, whereas subsidies draw down government budgetary revenues. This result helps explain why multiple intervention policies of the type described in Figure 3.6 are so rare, even though the benign transfer policies reduce efficiency losses in consumption or production. Policies that are inexpensive to administer might be chosen regardless of the magnitude of efficiency losses for consumers and producers. Therefore, policies that create separate producer and consumer prices are less common than policies that operate with a unified domestic market price.

Additional Efficiency Effects

The PAM provides only part of the information necessary for the evaluation of the efficiency effects of policy. Consumption efficiency losses are ignored; and except for direct subsidies to producers (SPI, SPE) and subsidies on inputs (S_I, S_E), policies for tradable commodities change prices to consumers and create consumption efficiency losses. All nontradable-commodity policies create consumption efficiency effects. Consumers gain when nontradable-commodity prices decline; they lose when these prices increase.

Another aspect of resource costs not included in the PAM is the administrative feasibility of policy instruments. Infeasible policies are usually policies that cost too much to administer. Trade restrictions are often the most cheaply administered policies; border controls can be established with only a modicum of difficulty, although smuggling vitiates total enforcement of policy. The success of subsidy policies depends on the ability of government agencies to distribute the subsidies to the groups targeted to receive them. Direct payments to farmers are straightforward in high-income economies with excellent communications and detailed records of output. But producer subsidies might be impossible in developing countries in which many small-scale farmers have little contact with the government. Policies creating price margins between consumers and producers that depart from normal marketing costs can be undercut by arbitrage. If transportation costs are small, if the form of the commodity is identical, and if law enforcement is poor, the same ton of a commodity might be recycled several times to receive a subsidy. Similarly, attempts to establish commodity taxes can be undermined by parallel markets that create a direct transaction linkage between producer and consumer. Determination of the optimal policy instrument is thus an empirical issue that can differ widely among countries, commodities, and government objectives. High implementation costs might easily outweigh consumer and producer efficiency losses.

A final category of resource costs results from lobbying government for the policy. Because inputs are used by producers or consumers in seeking policy-induced transfers, the economy forgoes some commodities or services. The costs of rent-seeking activity are often difficult to measure, because interest groups do not gain from revealing the degree of their efforts to receive favorable treatment from policy-makers. If bribery or other forms of corruption are entailed in the rent-seeking process, policy-makers will be similarly disinclined to reveal the extent of lobbying activity. But when the magnitude of potential transfers is large, at least the potential exists for a substantial allocation of resources to the lobbying effort.

Efficient Policies That Offset Market Failures

In the absence of market failures, all commodity price policies are distorting because they result in losses of efficiency for producers, consumers, or both. When product and factor markets operate efficiently and thus determine prices that reflect fundamental scarcities in an economy, no price policy can improve this already efficient outcome. But if some markets do not generate efficiency (social) prices, efficient government policy can intervene and correct the market failure. Ideally, the government should attempt to use the policy instrument that most directly offsets the divergence and thereby creates the largest efficiency gain.

Some of the most common efficient interventions are attempts by governments to control the movements in the domestic prices of principal food staples. In perfectly competitive conditions, insurance or futures markets are assumed present, so that producers can buy any desired amount of protection against instability. But in developing country agriculture, such markets almost never exist, and governments intervene directly to stabilize prices. Successful price stabilization can reduce price risks in agricultural production, guarantee stable markets, and obviate the need for costly adjustments by both producers and consumers to fluctuating prices and profits. These efficiency effects are also supplemented by the impact of price stabilization on nonefficiency objectives. Food price stabilization permits governments to maintain control of a critical parameter affecting the production decisions of farmers, the real incomes of urban consumers, and the nutritional status of the poorest people.

Indicators of success in stabilizing domestic food prices are sometimes available within the PAM framework, but only if the PAM data have been collected for a number of consecutive years. The purpose of stabilization is to hold the domestic price within some desired range irrespective of movements in the world price. Some countries attempt to follow expected world prices and to reduce domestic price variation relative to that found in the world market. Successful stabilization requires keeping the domestic price within the targeted range by supporting the floor price to producers through purchases of crops and holding the ceiling price to consumers through injections of food stocks. Performance can be gauged by the contrasting of movements in actual intrayear prices with those in the targeted range. Over time, the impact of the stabilization program can be found by comparison of the variation of domestic consumer and producer prices with that of world prices. Stabilizing effects of domestic stock policy can be identified if the variation of real (inflation-adjusted) domestic food prices is less than that of real international prices. Price stabilization policies entail costs for accumulation, storage, price monitoring, and distribution of a food commodity; and the calculation of these costs facilitates judgments of policy success.

Market failures can also be associated with commodities that create externalities, such as pollution and congestion. Such failures are often difficult to identify and measure accurately. Price policies sometimes over- or undercorrect for divergences because of these information problems or because the policies are aimed mainly at nonefficiency objectives. Only rarely are corrective policies targeted explicitly to particular market failures. As a practical matter, therefore, PAM analysts can only approximate adjustments in the net policy transfer (L) to show the effects of efficient policy. Under conditions of complete efficiency, all price policies would efficiently offset all market failures, and the residual distorting policy transfer (L) would be zero. But information sufficient to permit the realization of this ideal policy outcome almost never exists.

Concluding Comments

The thrust of policy analysis with the PAM approach is to identify the efficiency effects of a policy first and then to look for nonefficiency goals that might or might not justify incurring any efficiency losses associated with distorting policy. In the absence of market failures, all price policies create production or consumption efficiency losses, because the policies cause departures from optimal amounts of international trade. The most efficient levels of imports or exports are altered by policy, and the efficiency losses arise when too little or too much is traded. These trade effects of price policy are also of major interest to the country's international trading partners.

Analysts attempt to measure feasibility, implementation costs, budget transfers, efficiency losses, trade effects, and, finally, the impact on government objectives. In this way, the links between PAM budgets and price policy analysis can be drawn fully. However, policy analysis focused on a single commodity requires two critical extensions—examination of feedback in product and factor markets in the context of macroeconomic price and macroeconomic policy and consideration of the long-run dynamic effects of policies.

Bibliographical Note to Chapter 3

The analysis of commodity price policies is discussed in many economics textbooks, treatises, and articles, but none of these sources approaches it in the context of the policy analysis matrix. Three recent books contain especially useful bibliographical references that deal with price policies. The bibliographical note to chapter 4, "Marketing Functions, Markets, and Food Price Formation," of C. Peter Timmer, Walter P. Falcon, and Scott R. Pearson, *Food Policy Analysis* (Baltimore: Johns Hopkins University Press, 1983), pp. 150–214, summarizes principal contributions to the literature through 1982 on agricultural markets and prices and on the effects of policies influencing them. That listing is extended and updated in C. Peter Timmer, *Getting Prices Right: The Scope and Limits of Agricultural Price Policy* (Ithaca: Cornell University Press, 1986). A third review of the price policy literature, one not focusing on agriculture, appears in Anne O. Krueger, "Trade Policies in Developing Countries," in *Handbook of International Economics,* ed. Ronald W. Jones and Peter Kenen (Amsterdam: North-Holland, 1984), pp. 519–69.

Three textbooks offer especially useful adjuncts to the concepts developed in this chapter. William G. Tomek and Kenneth L. Robinson, *Agricultural Product Prices* (Ithaca: Cornell University Press, 1981), treats price formation and policy principally from the viewpoint of the agricultural sector of a single country; Richard E. Caves and Ronald W. Jones, *World Trade and Payments: An Introduction* (Boston: Little, Brown, 1981), discusses the theory of international price formation and of international trade policies; and Alex McCalla and Tim Josling, *Agricultural Policies and World Markets* (New York: Macmillan, 1985), analyzes the theory and application of domestic and international agricultural policies. These three texts also contain substantial citations.

The principles underlying the analysis of the effects of price policy on economic welfare can be found in the textbooks just cited. Detailed exposition of the concepts of producer and consumer surplus is provided in Richard Just, Darrell Hueth, and Andrew Schmitz, *Applied Welfare Economics and Public Policy* (Englewood Cliffs, N.J.: Prentice-Hall, 1982), chaps. 4–6. Extensions of these standard principles, with full references, appear in W. M. Corden, "The Normative Theory of International Trade," in Jones and Kenen, *Handbook of International Economics,* pp. 63–130; and in E. J. Mishan, *Introduction to Normative Economics* (New York: Oxford University Press, 1981).

A detailed analysis of price policies affecting rice in five countries is contained in Scott Pearson et al., *Rice in West Africa: Policy and Economics* (Stanford, Calif.: Stanford University Press, 1981); this book uses an approach that is an early variant of the PAM. An extensive PAM analysis of agricultural price policies affecting a wide variety of crops and livestock products in Portugal appears in Scott R. Pearson et al., *Portuguese Agriculture in Transition* (Ithaca: Cornell University Press, 1987). The literature on efficiency and price stabilization is vast; an empirical application of these concepts is provided in Cathy L. Jabara and Robert L. Thompson, "Agricultural Comparative Advantage under International Price Uncertainty: The Case of Senegal," *American Journal of Agricultural Economics* 62 (May 1980): 188–98. An empirical study of optimal policy choice that includes both economic welfare changes and implementation costs of trade and subsidy policies is reported in Eric Monke, "Tariffs, Implementation Costs and Optimal Policy Choice," *Weltwirtschaftliches Archiv* 119 (June 1983): 281–96.

CHAPTER 4

Factor Policy

FACTOR POLICIES are interventions that directly influence the prices of labor, capital, or land. Because a factor policy affects all commodity markets simultaneously, its impact on an economy can be more substantial than that of a commodity policy. This chapter reviews the principal causes of divergences in factor markets and explores how factor policies alter the profitabilities of particular production systems. These policies are considered first in a static context. However, a number of aspects of the process of income growth are missed by a comparative static view. Prices, factor endowments, and technologies change over time, causing changes in production patterns and incomes. Factor prices and factor policies have particular importance in the growth process because they influence the pattern of technical change as well as the profitability of commodity systems. Subsequent sections thus consider dynamic interactions among factor policy, public investment, and technical change. The final section places PAM results in the context of factor policy evaluation.

Classification of Factor Policies

If competitive market conditions prevailed, the PAM would use the same factor prices in evaluating private and social costs. But this coincidence is a rare event. Private market prices differ from their social values for a number of reasons. Commodity market distortions affect the factor markets indirectly, because changes in product prices alter the marginal value products of inputs. More important are direct govern-

57

ment interventions in the factor markets. Factor policies can be responses to dissatisfaction with the income distribution consequences of efficient outcomes. By altering the prices of factors, policy-makers hope to alter the share of total income received by the factor. Factor policies may also be a component of macroeconomic policy; for example, credit policies are used often to regulate the distribution of financial capital between the private and public sectors. Alternatively, private and social factor prices can diverge because of noncompetitive or natural phenomena, and efficient policies are introduced to establish a coincidence between private and social values.

Distorting Policies—Regulated Prices

The analysis of regulated prices needs to consider two general cases: "high" prices, in which regulations require increased payments to factors relative to their free-market levels; and "low" prices, in which existing factor prices are reduced relative to socially efficient values. Figure 4.1 considers the impact of a legislated increase in the payment to a particular factor. The legislation is assumed to provide minimum wages for unskilled labor. Initial equilibrium is determined by the intersection of the demand curve for labor, DD, and the supply curve for labor, SAS. If one assumes that no other divergences are present in the economy, the equilibrium wage rate, w^S, is the social value of labor. The government considers this wage rate too low and mandates an increase in the wage to w^P.

If the new wage rate can be enforced, producers will adopt it in their choice of inputs and use lesser amounts of labor that provide a higher marginal value product (equal to w^P). This employer response will result in a decline in total labor demand, from L^S to L^P. The policy will increase the income of employed workers, and it could have increased the total income of the labor force (if $w^P \times L^P$ had been larger than $w^S \times L^S$) and thus altered income distribution in favor of employed workers. The cost of the policy is represented by the unemployment created, which will be greater the more elastic the demand for unskilled labor. Additional costs of enforcement are incurred to ensure that employers comply with regulations. If these costs are so large that enforcement is not effective, private market wage rates will bear no relationship to w^P and instead will remain at or near w^S.

Most legislation to reduce factor prices applies to the capital market. Three motivations for controlling interest rates can be identified. First, low interest rates are seen as beneficial to economic growth because they

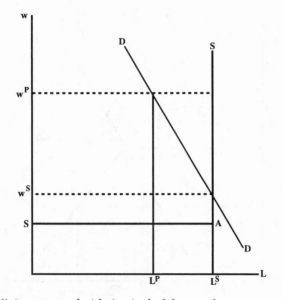

Figure 4.1. Minimum-wage legislation in the labor market

encourage increased investment. Second, controls on credit availability may be a principal instrument of monetary policy. By reducing aggregate demand through credit controls, the monetary authority hopes to limit inflation. In many cases, policy-makers' concerns that the limited volume of credit will become too costly result in the simultaneous establishment of low interest rates and credit rationing. Third, interest rates may be lowered to encourage investment and adoption of new technology by low-income borrowers, such as small farmers.

Figure 4.2 illustrates the impacts of controlled prices in the credit market. The demand for capital reflects the marginal value product of investment; the domestic supply of capital is provided by savings. The supply curve is drawn with a positive elasticity to indicate the ability of consumers to reduce current consumption levels as the reward to savings increases. Initial equilibrium is represented by price r^S and quantity K^S. If no divergences are present, r^S represents the social value of capital. A legislated reduction in the interest rate has two effects. The demand for capital increases to K_D^P, because more investments are possible when the cost of capital (the rate of return) is reduced. The supply of capital declines to K_S^P, because providers of capital are encouraged to favor current consumption over saving.

The lower interest rate creates an excess demand for credit, equal to $K_D^P - K_S^P$. The government has three options to resolve the excess de-

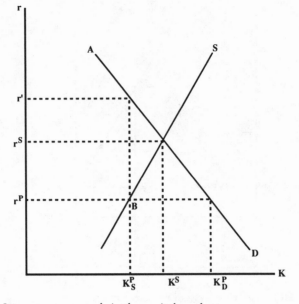

Figure 4.2. Interest-rate controls in the capital market

mand. First, it can choose not to enforce the regulated rate; investment levels and rates of return would then move back to their original equilibrium levels. A second option is to increase available supplies of credit for investment. This increase could be achieved by borrowing in external credit markets or receiving external aid, by channeling tax revenues into domestic credit markets, or by raising domestic interest rates to savers through government subsidy. When supplies are increased to meet all the demand for credit, the private market rate of return is r^P.

If the government cannot afford the cost of the savings subsidy or the budget is too constrained to shift financial resources into the credit program, a third option, rationing, becomes necessary. Available supplies of credit are limited to K_S^P, and the government introduces some mechanism to allocate the supply among users. In this circumstance, the private rate of return can be higher or lower than the social rate of return. The demand schedule shows that the marginal private rate of return to a total investment of K_S^P can be as high as r'. This rate would prevail if the rationing program allocated capital to uses that would provide the highest return.

Successful bribery could result in a similarly high private rate of return. The difference between the rate of return and the borrowing cost represents excess profit, and potential investors might be willing to use some of the excess to ensure access to credit. Because investors with the highest rate of return have the greatest ability to bribe, the marginal private rates of return may again be as high as r'. But if officials in charge of credit allocation are not corrupt, other allocation procedures will be used. The private rate of return may range anywhere from r^p to r'.

Legislation to reduce factor prices can also be found in the land rental market; in this case, low prices are intended to ensure that wealthy landowners will not exploit poor tenants. Analysis of regulations on land rental rates is analogous to that of regulations involving the capital market. As in Figure 4.2, the supply curve is portrayed as upward sloping. The supply of land to the rental market varies because rental rates affect the landowners' choices between self-cultivation and off-farm employment. Rental rates below equilibrium levels again create excess demands. The disequilibrium is resolved either by the creation of a parallel market, in which private rental rates rise above their social values, or by the forced exclusion of potential tenants, so that the private rental rate is held equal to the legislated price. In both cases, private prices diverge from their social values.

Distorting Policies—Taxes and Subsidies

A second group of distortions consists of employer taxes or subsidies on factors. When supplies of factors are inelastic, direct taxes on the owners of factors are relatively efficient ways to raise government revenue. If the supply of labor is perfectly inelastic, for example, personal income taxes can be levied without reducing the labor supply. But policy-makers are usually unable or unwilling to meet a large share of government revenue needs with direct factor income taxes; monitoring and enforcement costs are common constraints to increased tax collection. Another way to tax factors is to levy taxes on employers of factors. Perhaps the most common examples of these policies are social security taxes; both employers and employees are required to contribute to some benefit plan that provides unemployment insurance, health insurance, or retirement pensions. Because these taxes raise the cost of labor to the employer, employment is reduced until marginal value products equal the tax-inclusive wage rate. This result is illustrated in Figure 4.3a. Private market wage rates exceed social wage rates by the amount of the tax.

In some circumstances, however, the tax may be passed back to the factor, becoming similar to an income tax. If the supply of the factor is

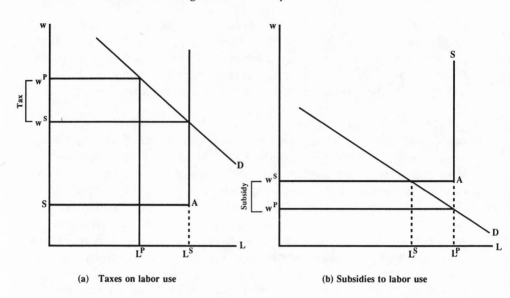

(a) Taxes on labor use   (b) Subsidies to labor use

Figure 4.3. Employer tax and subsidy programs in the labor market

perfectly inelastic, workers would be willing to perform the same jobs at wages lower than w^S; wages could decline as low as the wage represented by line segment SA (the subsistence wage) before labor supplies would decline. In this case, employer taxes are indirectly paid by the employee, because competition for employment causes money wages to decline by the amount of the tax. After the tax, employers still pay w^S in total compensation; private and social costs remain equal. Now, however, this amount goes to the government treasury. Full employment is maintained, but at a lower rate of employee remuneration (although employees may receive future benefits from these taxes in the form of retirement or insurance programs). Because the taxes are set in proportional terms, government enforcement will be unable to prevent deterioration in the wage paid to the factor; instead, enforcement only reinforces the decline.

Figure 4.3b illustrates the relationship between private and social values in the event of a factor subsidy. Demand for labor, as represented in the figure, is insufficient to absorb all supplies at the subsistence wage rate. A wage subsidy, equal to $w^S - w^P$ per unit, increases labor demand to meet full employment. This case reemphasizes the potential disparity between efficiency prices and optimal economic policy. Subsistence wages with full employment are a goal most governments desire. Because the social value of labor at full employment is only w^P, the policy-

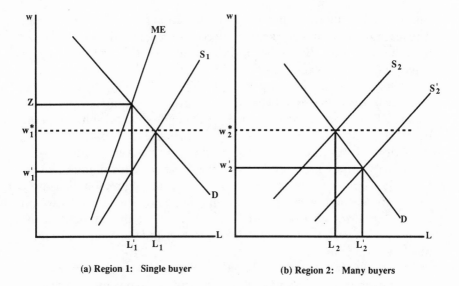

(a) Region 1:   Single buyer   (b) Region 2:   Many buyers

Figure 4.4. Monopsony and wage rates in the labor market

created divergence between private and social prices might be seen as desirable by policy-makers.

Efficient Policies—Offsetting Market Failures

Monopsony and oligopsony are common failures in factor markets, asserted to exist usually in the context of local labor markets. In this circumstance, only a single buyer (or group of colluding buyers) is bidding in the labor market. These market conditions provide the employer with an opportunity to pay labor a wage rate less than labor's marginal value product. Observed (private market) wage rates thus understate the social value of this labor.

Figure 4.4 provides a graphical portrait of a labor market for two regions. Initial equilibrium is w_1^*, and total labor supply is $L_1 + L_2$. The labor supply curve in each region is drawn with an upward slope. As wages in region 1 rise above w_1^* (caused, for example, by a rightward shift in labor demand), labor migrates from region 2 to region 1 and employment increases in region 1. This migration causes a leftward shift of the labor supply curve in region 2, which continues until the wage rates are equal in the two regions. Although total employment, $L_1 + L_2$, is fixed, the distribution of employment has shifted toward region 1.

When only a single buyer is present in one region, however, the supply curve is not the only relevant information for the hiring decision. The marginal cost of a worker is now considered, and this cost is larger than the wage rate. Hiring beyond L_1 requires that the employer pay a wage higher than w_1^* to all workers. Hence, the marginal expenditure on labor is larger than the wage rate at each level of employment. This relationship is represented for region 1 by the marginal expenditure curve, ME. The profit-maximizing level of employment results when the marginal value product, indicated by the demand for labor (D), equals the marginal expenditure on labor; this point is represented in the diagram by employment level L_1' and wage rate w_1'. An amount of labor, $L_1 - L_1'$, migrates from region 1 to region 2, shifting region 2's supply from S_2 to S_2' and lowering wages in that region to w_2'. The employer in region 1 realizes an excess profit of $L_1'(Z - w_1')$; this amount is the difference between the marginal value product and the wage rate, multiplied by total employment.

The results of calculations based on Figure 4.4 also show that the single buyer in one region is not sufficient to make monopsony an effective tool for exploiting the labor force. Either labor must be inflexible about moving between regions, or the region under monopsony must contain a substantial portion of the labor force, so that a reduction in wages within the region causes an outflow of labor sufficient to depress the wage rate in the other region. If, instead, region 1 were very small compared to region 2, attempts to force down region 1 wages would create an outflow of labor that would hardly affect wages in region 2. The supply of labor in region 1 would continue to decline (supply curve S_1 would shift leftward) until wage rates were equalized, in essence destroying the monopsony.

The second principal category of factor market failures consists of limits to the regional mobility of domestic factors. These constraints, usually termed institutional market failures, cause divergences between private and social prices regardless of the degree of competition among buyers of the factor input. In capital markets, rates of return to investment vary widely by region, sizes of investors, and production activity. Some of these differences exist because of deliberate policy-induced distortions. The market failures arise when institutions are absent because of government oversight or a lack of public investment funds to undertake necessary complementary investment. As a result, road and communication facilities are insufficient for banks and other financial institutions to enter particular regions. Insufficient legal codes of conduct or the absence of insurance provisions inhibit certain borrowing,

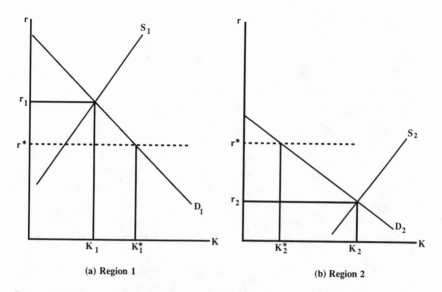

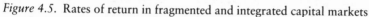

(a) Region 1   (b) Region 2

Figure 4.5. Rates of return in fragmented and integrated capital markets

lending, and savings transactions. These circumstances can cause unnecessary fragmentation in the capital market.

The consequences of fragmentation are illustrated in Figure 4.5,
which portrays capital markets fragmented between two geographical
regions. Other types of capital market fragmentation involve size of
borrower (small farmer versus large farmer) or type of productive
activity (agriculture versus industry). The demand curves for the two
regions, D_1 and D_2, portray the relationships between rates of return
and total investment (K). The supply curve represents the provision of
investment funds by savers plus the transaction costs necessary to move
funds from savers to borrowers. Region 1 is characterized by ample
investment opportunities and limited availability of funds relative to
region 2. As a result, the private interest rate in region 1 (r_1) is higher
than in region 2 (r_2). Because of the market failure, the private rate of
return in region 1 is above its social value; in region 2, the relationship
between private and social values is reversed. If the capital market were
integrated, supplies of capital would move freely from one region to the
other. Each borrower of capital would face a single supply schedule.
The market equilibrium and social rate of return would be r^*, found at
the intersection of $S_1 + S_2$ and $D_1 + D_2$ (not shown in Figure 4.5). Total
investment would increase in region 1 (K_1 to K_1^*) and decline in region 2
(K_2 to K_2^*).

Dynamic Effects of Factor Policies

Divergences in factor prices take on heightened importance in a dynamic context because of influences on the pattern of technological change. Technological change tends to reduce dependence on scarce, relatively expensive resources. For producers, private market prices rather than social prices are the indicators of relative scarcity, and divergences can allow technological choices that increase income in private market terms but reduce it in social terms. Similar possibilities arise in considering technological transfers, in which production practices are imported from foreign countries that have relative factor scarcities very different from those of the recipient country. The important divergences in a dynamic context may be different from those that dominate static analyses. Capital market distortions, for example, are usually unimportant in the evaluation of labor-intensive agricultural systems. But such policies can assume major importance for the process of structural change and the future development and adoption of more capital-intensive production techniques.

Public investment decisions are also critical to the process of technological change. The public sector is left with much of the responsibility for creating and introducing many of the innovations and for making investments complementary to technical change, because many of these activities involve public goods. Roads to market inputs and outputs, infrastructure for water delivery and electricity, institutional support for the extension of the financial system to rural areas, and education (schooling and extension) are examples of services that benefit agriculture but are not adequately provided by the private sector. Domestic research and development of new technology are rarely undertaken by the private sector, unless the technological change is embodied in a single input not easily replicated (for example, improved hybrid seeds). To the extent that public investment decisions are influenced by private rather than social incentives, factor and commodity divergences again influence the pattern of technological change and growth.

Technological Change

Technological change allows the minimum costs of production to decline when factor prices are held constant. The impact of technological change is illustrated in Figure 4.6. Given only two available inputs, labor (L) and capital (K), the production isoquant Q_{food} shows the

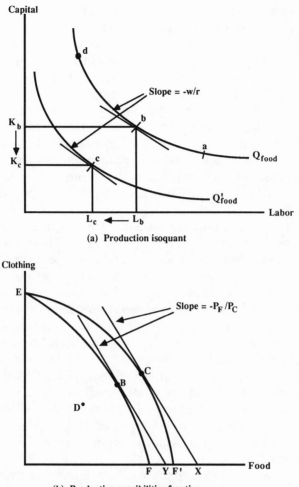

(a) Production isoquant

(b) Production possibilities frontier

Figure 4.6. Technological change, input substitution and divergences in factor prices

combinations of labor and capital inputs that can be used to produce one unit of food. To produce a given output, the producer considers the tradeoffs between additional capital input costs and increasing labor costs. In the figure, movement along the production isoquant from point a to point b raises capital costs by ($\Delta K \times r$) and lowers labor costs by ($\Delta L \times w$). When the producer finds the point where further changes in input combinations no longer reduce total costs—that is, ($\Delta K \times r$) = $-$ ($\Delta L \times w$), or $\Delta K / \Delta L = -w/r$—minimum costs of production are realized. The slope of the isoquant equals the negative of the relative price ratio. Point b represents this minimum-cost combination.

Technological change occurs with the introduction of a new production isoquant, Q'_{food}. Lesser amounts of both inputs are needed to produce a given level of output (K_b and L_b decline to K_c and L_c, respectively). The optimal input combination is represented by point c. Given a fixed supply of labor in the economy, the potential production of food increases. In Figure 4.6b, the technological change causes an outward shift in the production possibilities curve, from EF to EF'. Output increases from point B to point C; in this illustration, the output of both commodities increases. The total income gain to the economy can be measured in terms of either good; when food is the numeraire, the gain is $P_F(X - Y)$.

Analytical complications are introduced by the presence of divergences and by the possibility of input substitution. These complications are also illustrated in Figure 4.6a; if w and r represent the undistorted social prices of labor and capital, and if these factor prices are distorted by policy, so that labor becomes relatively more expensive and capital becomes relatively cheaper, the producer begins anew the search for the least-cost combination of inputs. Input substitution is practiced within the existing technology. Cheaper capital is substituted for more expensive labor, with the producer ending up at a point such as d. Relative to point b, labor input per unit of output has declined, just as in the technical change case. But the input of capital has increased. In aggregate terms, the economy shifts to an output point within the maximum production possibilities frontier (such as point D of Figure 4.6b) because the new technology provides an inefficient way to absorb total factor supplies.

The impact of factor distortions may be compounded over time as a result of the bias imparted to technical change. In the distorted situation, initial investment in research and complementary infrastructure begins with the firm and the economy located at points d and D in Figures 4.6a and 4.6b, respectively. If the investment is successful, the production isoquant will shift inward, at least with respect to point d. The production possibilities curve then shifts outward relative to point D. But unless the new production point shifts outside (northeast) of line YB, investment resources have been wasted; the investment was spent to move the economy to an income that could have been realized without any technological change (instead allowing producers to respond to social factor prices).

Further, the impacts of investment on the isoquant are not likely to be uniform; more likely the investment has greatest impact relative to the initial starting point and initial factor price incentives. An investment made at the factor prices corresponding to point d generates a new technology (and isoquant) that will look different from the isoquant portrayed as Q'_{food} in Figure 4.6a. Costs for capital-intensive techniques will have been reduced more than costs for labor-intensive techniques, and point C would not be on the new isoquant. In this circumstance, the production possibilities frontier will not reach surface ECF' in Figure 4.6b. The economy would have realized larger income growth by focusing investment on cost reductions for more labor-intensive techniques. Of course, non-efficiency objectives may be paramount to the investment decisions taken; the point here is that the efficiency costs of factor policy may well be larger than those suggested by partial equilibrium analyses of the type described above.

Dynamic Externalities

As industries mature and expand, they may generate new technologies or improved inputs that will benefit other industries. Such dynamic externality benefits, including improvements in labor skills and access to international markets, can arise from interactions across a wide range of industries. Alternatively, some particular subset of industries might be considered the prime generator of technical change and interindustry externalities, mandating an unbalanced approach to economic growth. Factor market policies and public investments are likely to play a critical role in this process (although commodity price policies can also be important), because they encourage the development and use of particular types of capital and labor inputs.

In agriculture, intercommodity externalities are relatively rare. Most technical changes are generated outside the production sector and are commodity-specific by necessity. Most dynamic externalities therefore arise from intraindustry effects. The infant industry argument is the most common example of this type of dynamic externality. Average costs are high relative to output price in the initial period. But over time, learning by doing causes costs to decline. Producers discover more efficient ways of operating, and labor productivity increases as workers develop better understanding of their jobs. In a future period, average cost is below the world price. The static perspective in the first period would judge the activity to be inefficient and socially unprofitable. In contrast, the dynamic perspective would see positive net benefits and a socially profitable activity. The short-run losses of the first period are a necessary consequence of the industry's operation, but the net present value of these losses plus the later gains could be positive.

In agriculture, the phenomenon of increased efficiency over time is commonly observed when a technology embodying new inputs is introduced. Farmers rarely know a priori the optimal amounts and timing of fertilizer application. But after several seasons of trial and error, physical relationships become better understood and efficient economic decisions can be made. Similar considerations apply to marketing and processing. How to operate new machinery, develop optimal methods of quality control, and tailor product specifications to the demands of particular consumer groups is rarely known perfectly in advance. For this reason, profitability estimates do not place much emphasis on initial experience.

Problems with the infant industry argument arise when gains in efficiency prove to be less dramatic than expected. Policies to support the adoption of fledgling technologies include direct producer subsidies, protection from imports (or subsidization of exports), and subsidization of inputs that embody the technology. But infant industry protection can create a situation of enduring positive private profitability (D > 0) and negative social profitability (H < 0). If efficiencies do not improve over time, the industry remains dependent on policy for its existence. Industry lobbyists then shift from infant industry arguments to other rationales in an attempt to maintain the support of policymakers for an inefficient activity.

PAM and the Evaluation of Factor Prices

Some factor policy issues can be addressed fully with PAM. Questions of policy bias toward a particular technology, for example, are addressed by comparisons of private and social profitabilities of alternative technologies for the production of a given commodity. The PAM results indicate the efficiency costs of technologies (if social profit, H, is less than zero or less than its value in alternative systems) as well as the price incentives necessary to elicit adoption of them (private profit, D, is positive and larger than the private profit for alternative systems). If D ≥ 0 and H < 0, policy encourages inefficient technical change; D < 0 and H ≥ 0 reflect conditions where policy discourages efficient technical change.

In most cases, tradeoffs and complementarities with nonefficiency objectives will be central to judgments about factor policies. Comparison of private and social values of factors in the PAM allows the analyst to contrast the consistency of income distribution objectives with production incentives. The impacts of policy on income distribution among factors are found by disaggregation of the factor cost element into returns to individual factors. The private cost of factors, C, can be divided into the returns to capital, C_K, to labor, C_L and to land, C_T; similarly, the social cost of factors can be disaggregated into G_K, G_L and G_T. Ratios of private and social factor cost can then be computed. If the system under study accounts for only a small share of total factor employment, it will not have much influence on income distribution. But if the system has positive private profitability, it can be considered consistent with the observed distribution of income.

Estimation of the net impact on employment requires additional information on the potential aggregate output or the number of producers represented by the system. To expand commodity system results to a national level, one selects estimates of potential aggregate output that conform to the unit of measure of the PAM. If the PAM results are measured in per hectare terms, for example, the analyst needs to estimate the potential area of the system and multiply it by the per hectare labor use to obtain total employment generated by the system. This result gives the gross employment impact of the system. The net contribution to employment then can be determined by comparison with employment declines in alternative commodity systems. The difference between the new aggregate employment of the expanding system under study and that of the contracting systems is the net effect on labor demand. This number will be positive if the expanding system is more labor-intensive than the contracting systems.

Evaluation of public investment policies can be made in concert with PAM. The extension of technical changes to new regions sometimes requires the provision of public goods. Adaptive research and modification of a commodity system will be necessary when agroclimatic conditions vary. Transportation, marketing, and information infrastructures might also be needed. Conceptually, the treatment of public goods in the context of PAM is straightforward. The analyst identifies investment policy costs that are complementary to the expansion of socially profitable systems. These costs are considered tradable-input costs (F) and domestic factor costs (G). The analyst is interested in the difference between additions to social profit (H) and public sector investment costs (F + G). If that difference is positive, efforts to support the expansion of the commodity system can be justified in efficiency terms. The costs (F + G) also represent the claim of the public sector on social profits.

Some public goods or services, however, cannot be attributed to particular commodity systems, because the product is indivisible over a wide variety of users. Road investments, for example, benefit consumers and nonagricultural producers as well as agricultural producers. An irrigation infrastructure is likely to benefit the producers of commodities besides those under study. Consumers vary in their willingness to pay for public goods, and commodity systems differ in social profitability and the capacity to pay. In such circumstances, the government can assess the true value of a public good only by aggregating willingness to pay across all users. PAMs then need to be prepared for all the commodity systems that benefit.

Because public investments often have long useful lives, the profitability of these investments will depend on expected future profit as well as present profit. Assessments of future profitabilities of commodity systems can be made by modification of observed PAMs for expected changes in factor prices, intermediate input prices, and output prices. However, speculations about future factor prices are usually more hazardous than are projections of future commodity prices. Even if policy-makers were confident about guessing future prices for most commodities, factor prices can be affected by changes in product mix or technology. Expected declines in some output prices need not lead to declines in returns to labor and capital if the product mix alters in favor of outputs that are relatively higher priced. Widespread technological changes can offset the often-observed tendencies for rates of return to capital to decline and for wage rates to increase. The availability of land as a residual factor means that at least some, if not all, of the change in profitability of agricultural products can be absorbed by changes in the price of land.

Concluding Comments

Classification of factor price policies is the first step in identification of government objectives for domestic factors. But these policies must also be analyzed in terms of their impact on profitability. Estimates of static profitability allow assessment of the consistency between factor and commodity policies. If factor policies cause commodity systems to become unprofitable, producer behavior may thwart intended objectives. Estimations of dynamic profitabilities allow policy-makers to consider changes in factor policies. If projected profitabilities increase over time, policy-makers have the option of decreasing subsidies (or increasing taxes) to the system. Particular nonefficiency objectives can be attained at lower efficiency costs. But if projected profitabilities decline over time, consideration must be given to making commodity-specific investments that can ameliorate the foreseen decline in efficiency.

Projections of social profit allow estimation of the changes in revenues (through yield increases) and in labor, capital, or tradable-input costs necessary to sustain or increase future social profitability relative to its current level. Agricultural scientists (or processing, industrial, or food engineers) can then help determine whether the required changes in the productivity of inputs and outputs can be realized by public investment in research and development. Public investment programs can begin to be formulated. If not, policy-makers need to reassess the contribution of the particular commodity system to their nonefficiency objectives and to explore the potentials for other commodities to contribute to those objectives at lower efficiency costs. In either circumstance, economic analysis plays a guiding role in the allocation of public investment in research and development.

Bibliographical Note to Chapter 4

The most intensive consideration of the political economy of factor prices has occurred in the capital market. A host of reasons explain the substantial divergences in the capital markets of many developing countries. Distortions that favor low interest rates have evolved partly from the early development strategies of industrialization and import substitution and partly from the motivations of well-intentioned policy-makers that were reinforced by the lobbying of rent-seeking interests. Such policies then prevent the elimination of market failures, particularly the development of financial institutions in rural areas. A concise discussion of this process is found in Hla Myint, "The Neoclassical Resurgence in Development Economics: Its Strength and Limitations," in *Pioneers in Development: Second Series,* ed. Gerald Meier (New York: Oxford University Press, 1987), pp. 107–36. A more detailed theoretical exposition is provided in Ronald I. McKinnon, *Money and Capital in Economic Development* (Washington, D.C.: Brookings Institution, 1973). Much of the empirical work in this area is contained or referenced in three volumes: John Howell, ed., *Borrowers and Lenders: Rural Financial Markets and Institutions in Developing Countries* (London: Overseas Development Institute, 1980); J. D. von Pischke, Dale W. Adams, and Gordon Donald, eds., *Rural Financial Markets in Developing Countries: Their Use and Abuse* (Baltimore: Johns Hopkins University Press, 1983); and Dale W. Adams, Douglas H. Graham, and J. D. von Pischke, eds., *Undermining Rural Development with Cheap Credit* (Boulder, Colo.: Westview Press, 1984).

Analyses of land and labor market policy in agriculture have been more limited. A survey of the latter factor is provided in Lyn Squire, *Employment Policy in Developing Countries: A Survey of Issues and Evidence* (New York: Oxford University Press, 1981). Perhaps the best-known contribution in this area is the induced migration model of John R. Harris and Michael Todaro, "Migration, Unemployment, and Development: A Two Sector Analysis," *American Economic Review* 60 (March 1970): 126–42. Most recent research has focused on understanding the ways that land and labor markets function and the role of market failures. Two recent works are Hans Binswanger and Mark Rosenzweig, eds., *Contractual Arrangements, Employment, and Wages in Rural Labor Markets in Asia* (New Haven, Conn.: Yale University Press, 1983); and Pranab Bardhan, *Land, Labor and Rural Poverty* (New York: Columbia University Press, 1984).

The influence of factor prices in the process of growth and technical change is of prime importance in the induced innovation model of Yujiro Hayami and Vernon Ruttan, *Agricultural Development: An International Perspective* (Baltimore: Johns Hopkins University Press, 1971). This model is elaborated and extended to incorporate institutional development in Hans Binswanger and Vernon Ruttan, *Induced Innovation: Technology, Institutions and Development* (Baltimore: Johns Hopkins University Press, 1978).

More elaborate models of the role of policy and dynamic change have been sparse. One such model is developed in Michael Bruno, "Development Policy and Dynamic Comparative Advantage," in *The Technology Factor in International Trade*, ed. Raymond Vernon (New York: National Bureau of Economic Research, 1970), pp. 27–64. The comments following this article, by Robert Aliber and Nathan Rosenberg (pp. 65–72), elaborate on many of the reasons for these problems. An example of the projection of future comparative advantage, based on expectations about factor and output prices, is provided in Scott Pearson et al., *Portuguese Agriculture in Transition* (Ithaca: Cornell University Press, 1987).

CHAPTER 5

Macroeconomic Policy

MACROECONOMIC POLICIES comprise fiscal and monetary policy, budgetary policy, and policies that govern the economywide or macro prices—the exchange rate, the interest rate, and the wage rate. In most developing countries, macroeconomic policies have a major impact on the profitability of agricultural systems and the welfare of farmers. Governments typically extract a greater amount of tax revenue from agriculture than they spend on agricultural subsidies or investments. This bias against agriculture in budgetary allocations is then complemented with a pervasive tax on farmers, levied, sometimes unintentionally, through the exchange rate by skewed macroeconomic management. As a result, attempts to provide positive incentives to agriculture with commodity policies can be overwhelmed by negating macroeconomic policy that transfers resources away from agriculture and the rural economy.

This chapter focuses on these two primary biases of macro policy against the agricultural economy. The first part of the chapter introduces the central elements of budgetary, fiscal, and exchange-rate policies and examines how these dimensions of macroeconomic management can lead to biases in resource allocation to agriculture. The second part of the chapter is concerned with how the management of macro policy can impose a widespread tax on agricultural producers. This organization allows a clear focus on the two most critical macro-micro linkages.

Elements of Macroeconomic Policy

Macroeconomic policy has its most direct influence on agricultural profitability through decisions to collect and spend government budget-

75

ary resources. For agricultural systems, the implications of agriculture's share of government spending are clear. The indirect effects on agriculture of central government decisions to finance spending programs are more complicated. But an understanding of the relationships among fiscal policy, inflationary pressure, exchange-rate options, and agricultural profitability is critical. That set of relationships underlies the indirect imposition of a tax on most agricultural producers.

Budgetary Policy

Budgetary policy deals with the allocation of total revenue, both between recurrent and capital expenditures and among sectors of the economy. The link between budgetary policy and agricultural policy is straightforward; budgetary decisions constrain the levels of government resources available for agricultural programs, such as public investment or recurrent subsidization of agricultural production or marketing. The agricultural sector is only one of many recipients of government funds and, in most developing countries, absorbs only a minor share of such funds. Other categories of expenditure—military and defense, welfare programs for disadvantaged consumers, education and health investments, public sector industries, and public sector employment—account for much larger shares of the budget. Like agriculture, these categories of expenditure are also represented by interest groups with sets of objectives and desires for policies. These objectives often require budgetary support, and budgetary allocations thus serve as indicators of policy-makers' priorities among the competing sectors. But because some objectives can be served by policy instruments that impose little burden on the budget, the expenditure pattern reveals only part of the preference structure of policy-makers.

Fiscal Policy

Fiscal policy is the set of decisions that determine how much revenue the government collects. Ultimately, the government collects as much revenue as it requires for budgetary policy. But for analytical purposes, categorization of sources of revenue allows the identification of budgets that are deficit, surplus, or balanced. Recurrent revenues come from government income from taxes, the profits of public sector industry, and borrowings to finance productive investment. When these revenues are compared to recurrent budgetary expenditures, the difference is the budget surplus (revenues exceeding expenditures) or deficit (revenues less than expenditures). Such calculations have analytical importance because they reveal the accommodating policy actions that governments take to resolve budget imbalances. If the budget is in deficit, some form of additional borrowing must occur. This borrowing can be from domestic or foreign lenders, government reserves, or the government's central bank. If the budget is in surplus, the government has to dispose of the surplus. Options include increases in lending to domestic or foreign borrowers, increases in government reserves, or retirement of debts to the central bank or foreign lenders.

Calculation of the budget balance between recurrent revenues and expenditures is complicated by the distinction between productive and unproductive public investments and variations over time in economic activity and in public sector investment. Only rarely will the budget be exactly balanced. In a dynamic, growing economy, the government can maintain a policy of deficit spending that involves growth over time in total indebtedness. If the economy and the government's tax revenue are expected to grow in the future, the government can borrow in the present against that expected future income. When the expected future income gains are realized, the government repays the debt. In an economy that is growing continuously, this exercise is repeated regularly, allowing the government to operate with a deficit in every year. But unless substantial growth in tax revenues is foreseen, the magnitude of allowable deficits will be small. The concept of balanced budgets thus remains a useful (though approximate) basis for evaluation of macroeconomic policy.

Exchange-Rate Policy

A government can choose among three alternatives in establishing its regime to manage the exchange rate (units of domestic currency per unit of foreign currency). The alternatives differ mainly according to the degree of government intervention in the foreign-exchange market. This market exists because local export suppliers and foreign investors have foreign currency that they wish to exchange for domestic currency and local consumers desiring to buy imports or to invest abroad want to make the opposite currency trade. In the absence of government intervention, a market-clearing price for the exchange rate would be determined at a level that would balance the national supply of foreign exchange with the total demand for it.

The first type of exchange-rate regime is termed floating, or freely fluctuating. Since mid-1974, most Western industrialized countries have had floating exchange rates (although governments often intervene in the foreign-exchange markets by buying or selling on government account, a process that has led to the term *dirty floats*). When a government permits the foreign-exchange market to determine the exchange rate, this market-determined price fluctuates in line with underlying shifts in the supply or demand of foreign exchange. Therefore, a floating exchange rate forces adjustments in levels of imports, exports, and international capital movements.

The second kind of exchange-rate regime encompasses a maximal degree of government intervention in the foreign-exchange market. Governments of most developing and centrally planned economies use fixed (pegged) exchange-rate regimes, in which they determine the nominal value of their exchange rates. A fixed exchange rate is intended to shield the economy from market-induced fluctuations and to permit greater domestic use of macroeconomic policy. To establish a fixed exchange rate, a government first has to decide how to tie (or peg) its currency. It can tie either to one foreign currency, typically the one in which most of its foreign-exchange transactions take place, or to a basket of currencies, often weighted according to the importance of the included foreign currencies in the country's international transactions. Because the United States has a large role in world trade and capital flows, and because the transactions in some commodity markets (notably crude petroleum) are carried out entirely in U.S. dollars, the dollar is the currency most often chosen as the reference point for fixed-rate regimes. The dollar also weighs heavily in most currency baskets for countries that select this method of fixing their exchange rates.

A fixed-exchange-rate regime does not totally insulate the country's economy from market forces. If particular fixed exchange rates do not correspond to a balance of demand and supply for foreign exchange, additional government actions are required to address the imbalance. Rationing of foreign exchange, multiple exchange-rate regimes, and restrictions on certain categories of imports are common responses to an imbalance of foreign exchange. Tradable commodities subject to exchange rationing become nontraded; their price is determined entirely by domestic demand and supply conditions and will only by chance be as low as world prices. World prices remain relevant only for high-priority imports not subject to rationing.

One policy adjustment to exchange imbalance is revaluation. Under conditions of excess supplies of foreign exchange, revaluation decreases the amount of domestic currency needed to purchase a unit of foreign currency. This adjustment is an upward valuation of the domestic

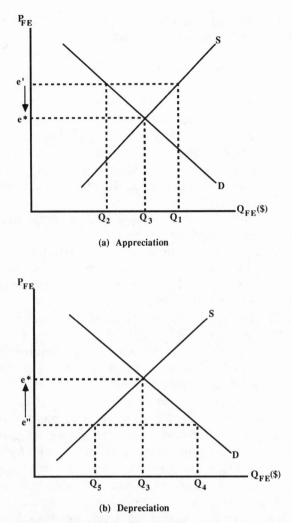

(a) Appreciation

(b) Depreciation

Figure 5.1. Adjustments in fixed exchange rates

currency, so that it appreciates relative to the foreign currency. The effects of appreciation are illustrated in Figure 5.1a. For convenience, foreign transactions are assumed to be limited to exports or imports of commodities. The supply curve of foreign exchange, S, represents the amount of foreign currency earned from sales of exports. That curve is upward sloping; as the price of foreign exchange (the exchange rate) rises, domestic producers of exportables find foreign market opportunities increasingly attractive. The demand curve, D, represents the amount of foreign exchange used to buy imports. The curve is downward sloping; because importables become cheaper in domestic prices as the exchange rate declines, consumers demand increasing amounts of importables.

At the initial exchange rate, e', the economy experiences a surplus of foreign exchange. Domestic producers earn Q_1 dollars, but domestic consumers spend only Q_2 dollars. The difference is absorbed by the government in the form of increased foreign-exchange reserves. This happens if the government needs to rebuild depleted reserves or to retire foreign debt. Eventually, the needs are satisfied, and continued accumulations are no longer desired. Appreciation of the currency, from e' to e^*, is one way to prevent continued accumulation of these reserves. Consumers increase their demand for dollars to Q_3; producers reduce their supplies of dollar-generating exports to the same amount. The exchange rate is now in equilibrium.

Excess demand is a more common condition in the foreign-exchange markets of developing countries. In these circumstances, a devaluation can address the imbalance. A devaluation is a policy decision to change a fixed exchange rate so that the domestic currency depreciates relative to foreign currencies (that is, more local currency is needed to purchase a unit of foreign currency). For example, if a fixed rate of 50 pesos per dollar is changed to 75 pesos per dollar, the devaluation of the peso results in a 50 percent depreciation of the peso: $(75/50 - 1)$ x 100 percent. Devaluations in fixed-exchange-rate regimes are usually catch-up actions. In effect, they are discrete changes that offset the cumulated effects of postponing gradual adjustment in the exchange rate.

The effects of carrying out a devaluation are shown in Figure 5.1b. At e'', excess demand for foreign exchange exists. Demand for imports is equal to Q_4 dollars, whereas exports are providing only Q_5 dollars. The government could meet the excess demand by reducing its reserves of foreign exchange. Otherwise, it would have to impose a rationing scheme. Following the devaluation, the world prices in domestic currency are raised by the percentage that the currency is depreciated. Local producers respond to the increased prices by expanding their output of importables and exportables, and local consumers react to the higher prices by reducing demand for these commodities. Import expenditures decline, from Q_4 to Q_3, and export earnings rise, from Q_5 to Q_3. Equilibrium in the balance of payments is restored.

Under freely floating rates, the adjustments described in Figure 5.1 occur automatically. Under fixed-exchange-rate regimes, adjustments take place only at the discretion of the government. A third type of regime for exchange-rate policy—a crawling-peg (adjustable-peg) re-ime—is intermediate between the first two and incorporates features of both. This regime is an attempt to match the control features of the fixed-rate regime with the market determination of the floating-rate regime. The crawling peg is a fixed-rate system in which the government announces in advance a schedule of weekly or monthly changes in the rate. These changes are meant to track the expected movements in the market—those that would have resulted if the rate had been allowed to float. Design issues in a successful crawling-peg regime include proper definition of a basket of currencies, usually trade-weighted, that reflects the country's main international transactions and appropriate choice of the schedule to change the rate (to approximate the movements in market-determined levels). Well-designed crawling-peg regimes provide developing countries with the policy and antifluctuation controls inherent in the fixed-rate system and the market-clearing adjustments provided automatically in the floating-rate system.

Macro-Micro Linkages

The impact of macroeconomic policy on microeconomic incentives is represented by two main linkages. The first is the impact of fiscal policy on the macro prices. Changes in these prices create changes in the prices of inputs and outputs and thus influence agricultural profitability. The second interaction is direct manipulation of the macro prices through government macro policy. Such action intentionally distorts factor prices or exchange rates, usually in the pursuit of nonefficiency objectives.

Fiscal Policy Linkages

Fiscal policy is of principal interest when the government budget balance is considered to be unsustainable. If budget balances fluctuate between surpluses and deficits but equal zero on average, macroeconomic policies represent stabilizing actions, such as the accumulation or distribution of government reserves. But if the budget balance is chronically in surplus or deficit, consumers and producers are affected directly.

The relationship between the government and the private sector can be demonstrated with the aid of the national income accounting identity. In the long run, the aggregate output available (the supply of goods and services) must equal aggregate expenditures (the demand for goods and services), as shown in the following identity:

Aggregate output   = Aggregate expenditures
(origin of income)   (uses of income)
$$Y = P \times Q + (F^{in} - F^{out}) = C + I + G \tag{1}$$
where
  Y = Monetary value of national output or income
  $P \times Q$ = Value of all goods and services produced
    F^{in} = Total value of foreign exchange inflows (exports and foreign capital inflows)
   F^{out} = Total value of foreign exchange outflows (imports and domestic capital outflows)
    C = Consumption expenditures in private sector
    I = Domestic investment expenditures in private sector
    G = Government expenditures on public consumption and public investment

Because $(P \times Q - C)$ equals domestic savings (S) plus taxes (T), equation 1 can be rewritten:

$$S + T + (F^{in} - F^{out}) = I + G$$
$$(S - I) + (F^{in} - F^{out}) = G - T \tag{2}$$

Equation 2 summarizes the necessary interaction between the budget balance $G - T$ and the rest of the economy. Positive values of $G - T$ correspond to a government budget deficit. In this circumstance, the economy must experience a surplus of domestic savings over domestic investment, a surplus of foreign exchange inflows over foreign exchange outflows, or some combination of the two. Because this case characterizes fiscal policy in most developing countries, the subsequent discussion will focus exclusively on fiscal deficits.

Several macroeconomic policy alternatives are available to generate the necessary balance in government resource use. Eventually, the government must alter budgetary policy and reduce expenditures (causing G to decrease) or increase taxation and transfer expenditures from the private to the public sector (causing T to increase). But this alternative is often limited in the short run because of political constraints. Increased tax collections may also be difficult to administer.

Alternatively, the government can try to borrow domestically, either by increasing the interest rate to decrease private consumption and

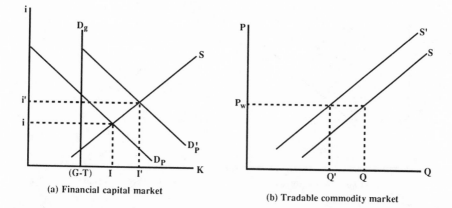

Figure 5.2. Fiscal policy, the capital market, and commodity market effects

private investment in favor of saving or by imposing limits on private investment. In both circumstances, S − I will increase. Figure 5.2 illustrates the effect of this macroeconomic policy on both the capital market and a typical commodity market. The financial-capital market is portrayed in Figure 5.2a. Initially, all borrowing is done by the private sector; investment is I, the interest rate is i, the intersection between the supply of savings, S, and the demand for investment, D_P. If the government decides to finance the budget deficit by borrowing, a government demand for capital, D_g, is introduced. Demand is perfectly inelastic at the amount of financial capital needed to meet the deficit (G − T). The new market demand curve is now D_P', equal to D_g plus the private sector demand curve (D_P). The interest rate rises from i to i′, and the private sector investment declines from I to I′ − (G − T) . Government policy has raised the cost of capital and the rate of return in the private sector. The effect on the commodity market is captured on the supply side, because rising interest rates cause the supply curve to shift upward, from S to S′. Production declines from Q to Q′.

Another policy option is to finance the public debt by borrowing on foreign capital markets, thereby increasing foreign capital inflows (permitting $F^{in} − F^{out}$ to rise). This option is governed by the country's ability to service foreign debt, itself a function of past foreign borrowing and future development prospects. Figure 5.3 more elaborately portrays the foreign-exchange market by recognizing foreign capital flows as well as trade in goods and services. The diagram for foreign capital flows, Figure 5.3b, is presented in terms of relative rates of return and amounts of foreign exchange. The supply curve for foreign exchange in

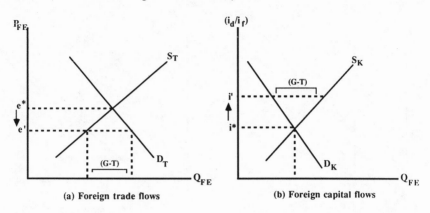

(a) Foreign trade flows   (b) Foreign capital flows

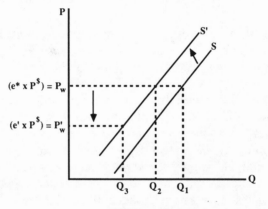

(c) Commodity market

Figure 5.3. Fiscal policy, the foreign-exchange market, and commodity market effects

the domestic economy, S_K, is upward sloping. Increased inflows result as the domestic rate of return (i_d) rises relative to the foreign rate of return (i_f). The demand curve, D_K, slopes downward, because as the domestic rate of return falls, investors want more foreign exchange (to use in foreign rather than domestic investments).

Initial equilibrium is represented by exchange rate e^* and interest rate i^*. Next a government deficit of $G - T$ is introduced, financed by the borrowing of foreign exchange. Now the offsetting surplus appears in the foreign capital accounts. Foreigners supply more capital, and domestic capital formerly destined for foreign countries is diverted to domestic markets. For the balance of payments to be maintained, foreign trade flows must generate a deficit to offset the surplus in the foreign capital account. This result is achieved by an appreciation of the exchange rate; the price of foreign exchange falls from e^* to e'.

Commodity producers are affected in two ways. If the foreign capital market is linked with the domestic capital market, interest rates will increase and cause the costs of production to increase, shifting the supply curve from S to S' and reducing output from Q_1 to Q_2. Moreover, the decline in the exchange rate will cause the domestic currency value of output to decline from $P^w(e^* \times P^*)$, where P^* is the foreign currency price of the commodity, to $P^{w'}(e' \times P^*)$. The output effect of this price change is represented by the movement from Q_2 to Q_3.

In many developing countries, foreign capital flows are tightly controlled and isolated from domestic capital markets. Interest-rate effects on the domestic capital market thus could be absent. However, the exchange-rate effects remain. The government deficit still augments total foreign-exchange supplies, and therefore the domestic currency price of foreign exchange is lower than it would be without the fiscal deficit.

The final option for financing deficits involves the use of monetary policy. This policy action usually entails borrowing from the country's central bank, which finances the government debt by increasing the money supply. If the country has unemployed resources that can be mixed together in proportions fitting existing technologies and management, the increase in the money supply can result in the growth of production of goods and services (Q will rise). Taxes and savings increase, and the deficit is financed. But most developing countries face binding resource constraints that cannot be broken simply by increased public expenditure. What is adjusted is the average level of prices (P inflates).

A deficit financed by monetary policy thus causes demand-pull inflation. As equation 2 shows, this policy cannot successfully finance the deficit unless savings increase relative to investment or net foreign inflows increase. In general, therefore, monetary policy actions must be accompanied by one or more of the options just described. The only other way for monetary policy to work involves money illusion; private consumption is reduced in real terms, allowing public consumption (financed by the increase in the money supply) to expand. Without money illusion, monetary expansion will cause inflation. Private consumers alter the prices of their services to maintain their real consumption levels; governments must borrow increasing (nominal) amounts of money to finance the increased (nominal) costs of the public sector demand for goods and services entailed in the budget deficit.

Exchange-Rate Policy Linkages

Governments also attempt to manipulate macro prices as an explicit objective of policy. Attention here focuses on the remaining macro price, the foreign-exchange rate. When exchange rates are fixed, distortions may be present even without fiscal imbalance. Figure 5.4 illustrates the case of an overvalued exchange rate, e'. The gap between the demand for and the supply of foreign exchange can be sustained as long as the government can borrow abroad or draw down its reserves of foreign exchange. But after these opportunities are exhausted, the country can continue to consume more tradables than it produces only if its trading partners agree to hold its currency, an unlikely occurrence for most developing countries.

If the government wishes to sustain a rate of e', it can try to induce shifts in the demand and supply curves. Curves D' and S' represent shifts that would enable the exchange rate e' to be sustained. These shifts could be attained in several ways. First, foreign income growth affects domestic commodity export opportunities as well as the financial-capital account. Second, shifts in demand and supply conditions in particular commodity markets could increase the world prices of exportables or decrease the prices of importables. Technology provides a third source of change in the foreign-exchange market. Widespread technological changes decrease production costs in the domestic economy, shifting outward the supply of exports and shifting backward the demand for imports. By raising domestic rates of return, domestic technological change also affects the foreign capital account. Inflows of foreign funds are encouraged, and outflows of domestic funds are discouraged.

Two of the adjustment options—foreign income growth and changes in world prices—are beyond the control of the government. Furthermore, changes in world prices would have to be widespread or involve commodities with prominent roles in the domestic country's foreign trade to have a substantial impact on the current-account position. Technological change, a more attractive option, generally requires substantial periods of time and amounts of investment. In most situations, therefore, the overvalued exchange rate must be complemented by additional distorting policies. These policies include rationing—direct allocations of foreign exchange (amount Q') to particular import uses—and protection of import-competing industries so that the demand curve for foreign exchange shifts leftward (intersecting S at e' and Q').

Because protection and rationing are not provided uniformly to import-competing industries, some sectors benefit from this combination

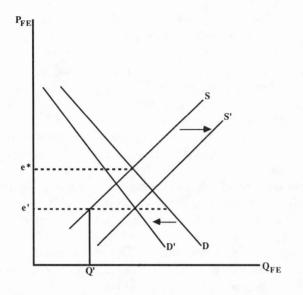

Figure 5.4. Exchange-rate distortions

of policies and others lose. Consequently, interest groups are likely to play a major role in determining the incidence of such protection. The degree of political power of various groups differs among countries. But a plausible generalization is that large-scale industrialists, whether local or foreign, have better access to political privileges than do agricultural producers. Like farmers and small-scale firms in developing countries, most large-scale industries produce goods that are tradable internationally. The production of manufactured goods would be taxed by macroeconomic policy that creates an overvalued exchange rate. If so, the owners, managers, and employees of large-scale urban industries could be expected to support devaluation. In this sense, their political and economic interests would be aligned with those of farmers and artisans who produce tradable goods in rural areas. All of these groups would wish to rid themselves of the exchange-rate tax caused by overvaluation.

Nevertheless, this political alignment of rural and urban producers rarely occurs. Instead, the politically well-organized urban producer interest groups are able to protect themselves by convincing the government to enact commodity policies that offset the taxing effects of the overvalued exchange rate. In the choice of a particular instrument to provide protection, there is an important difference between a tariff and a quantitative restriction. A tariff provides a one-time increase in the domestic price, since the tariff is calculated as a percentage of the cif import price. This world price (in domestic currency) is held constant by a fixed exchange rate. If the exchange rate becomes more distorted over time (the degree of overvaluation increases), the tariff will have to be increased accordingly to continue to offset the taxing effect of macro price policy. But a quantitative restriction provides continuing protection automatically. In this circumstance, the prices of tradable goods protected by quantitative restrictions respond fully to domestic inflation. Since only a limited amount of imports are permitted by the quota, the prices of quota-protected tradables are set by domestic demand and supply conditions.

It is thus no coincidence that special interest groups lobby to receive quantitative restrictions to offset overvaluation. In many developing countries, import quotas on favored goods are set at zero, providing complete protection to local industry. Once protected by import quotas, the benefactors are no longer indifferent to the issue of macroeconomic reform. They then have a strong incentive to oppose devaluation in order to keep the prices of tradable inputs from rising. The combination of quantitative restrictions and an overvalued currency thus shifts the politics of macroeconomic reform by moving urban industrialists from the supporting to the opposing side. Industrial entrepreneurs seek principally to insulate their firms from overvaluation and to receive special policy favors rather than to minimize production costs. Rural producers and macroeconomic reformers then face the imposing task of overturning these entrenched interests.

PAM and the Evaluation of Macroeconomic Policies

The evaluation of macroeconomic policy will usually take the analyst well beyond PAM's focus on production efficiency and relative profitabilities. Budgetary policy issues, the structure and feasibility of various tax collection systems, and approaches to monetary management and financial regulation are each entire disciplines. But as in the consideration of commodity policy and factor policy, the PAM can provide useful insights into some of the critical aspects of macroeconomic policy.

Fiscal Balance and Revenue Generation

Taxes create potentially large efficiency losses in production, and the PAM approach allows a focus on the revenue-efficiency tradeoffs that result. Net impacts on income are represented by H. The fiscal contribution of a system requires consideration of the specific instruments used to effect policy transfers (the third line of the matrix). Trade taxes or subsidies, for example, affect government revenue only when exports or imports are involved. Domestic production of these commodities generates no revenue.

The PAM results are also helpful in identifying ways to generate revenue more efficiently. The social profitabilities of various commodities provide an indication of a feasible set of taxes on a fixed input, such as land. The producer will have no incentive to shift out of a particular commodity so long as its profitability (D) remains positive and the commodity retains its ranking among alternative crops. A proportional profits tax meets these requirements. If there are no divergences in the output or input markets, the PAM matrix for a particular system becomes the following.

	Revenues	Tradable inputs	Factor inputs	Gross profit	After-tax profit
Private prices	A	B	C	D	$D - kH$
Social prices	E	F	G	H	H
Difference between private and social prices	O	O	O	O	$-kH$

Comparison of the social profitabilities of alternative crops in a region gives information about the potential for various levels of land taxes. Systems with higher social profits, ceteris paribus, will generate larger contributions to government revenue. As H increases, so does kH. But in general, the estimation of revenue effects requires aggregation of representative systems to the national output level. Small tax revenues per unit of output can become relatively important sources of aggregate revenue if total output is large. Subsequent calculation of the impact of taxes on private profits gives insights into whether commodity production patterns will be consistent with taxes on social profits. Because other divergences will cause private profits to differ from social profit, socially efficient taxes may be infeasible without the changes in the incentives given to alternative commodity systems.

Foreign-Exchange Balance

Governments that consider their short-run needs for foreign exchange to be inelastic have an interest in examining the direct foreign-exchange burden of particular agricultural systems. This burden is measured in social prices, based on world prices for inputs and outputs. Private prices are not relevant for this calculation, since they include the impact of domestic taxes or subsidies and thus do not reflect full foreign-exchange value. If the product is exported, the direct foreign-exchange impact (in PAM notation) is equal to $E - F'$, where F' represents tradable inputs that are actually imported. But foreign-exchange impacts can also occur indirectly, because domestic production substitutes for imports. If the country is a net importer of its staple food, for example, domestic staple food production earns no foreign exchange directly. But by being substituted for imports of the commodity, domestic production saves foreign exchange. A similar argument applies to inputs that are produced and used domestically but are potentially bought or sold on international markets. Consequently, a full measure of short-run foreign-exchange impacts is value added at world prices $(E - F)$.

The value of $E - F$ gives the immediate foreign-exchange impact of particular systems. But this measure is relevant only for a time period in which the domestic factors of labor and capital are immobile. In the long run, the social profitability of systems measures their impact on foreign-exchange supplies. The use of domestic factors by a particular system requires their withdrawal from some other production activity that can directly earn or save foreign exchange. The social value of domestic factors can thus be interpreted as an opportunity cost, measured in terms of potential foreign-exchange earning power. Allocating capital to the domestic production of rice, for example, denies the use of that same capital for the production of some other export or import-substituting commodity. Analogous reasoning applies to nontradable outputs. Ultimately, these products substitute at the consumer level for tradable goods. So long as consumers spend all of their incomes, withdrawal of a nontradable good is associated with increased consumption of some other good. Therefore, the full measure of the foreign-exchange effect is $E - F - G = H$; the long-run foreign-exchange impact of an agricultural system is identical to its social profitability.

Policy-makers do not necessarily face difficult tradeoffs in comparing long- and short-run availabilities of foreign exchange. A ranking of systems in terms of $E - F$ is likely to be quite different from a ranking of systems according to $E - F - G = H$. But a shift toward policies that promote socially profitable systems need not have adverse short-run impacts on availabilities of foreign exchange. Because resources move out of relatively inefficient systems (high values of $E - F$ but low values of H) into more efficient systems (high values of H), these shifts in

Table 5.1. The Foreign-Exchange Effects of Efficient Policies

	Revenue	Tradable inputs	Domestic factors	Profit
Initial Policy: Subsidize System 1, Tax System 2				
System 1				
Private prices	25	10	10	5
Social prices	25	10	20	−5
Policy effects	0	0	−10	10
System 2				
Private prices	100	60	60	−20
Social prices	100	60	20	20
Policy effects	0	0	40	−40
Policy Change: Eliminate Taxes and Subsidies				
System 1				
Private prices	25	10	20	−5
Social prices	25	10	20	−5
Policy effects	0	0	0	0
System 2				
Private prices	100	60	20	40
Social prices	100	60	20	40
Policy effects	0	0	0	0

production patterns generally increase the net supply of foreign exchange. This effect on foreign-exchange supplies is immediate, because both the direct and indirect impacts on foreign exchange occur instantaneously. A given quantity of domestic factor inputs earns more foreign exchange in efficient systems than in inefficient systems. The net foreign-exchange supply will be adversely affected only if domestic resources are underused during the transition from inefficient to efficient production. Hence, foreign-exchange goals are fully compatible with efficiency maximization.

These results are illustrated in Table 5.1. The country has two ways of producing a commodity, represented by system 1 and system 2. System 1 offers a value-added in domestic prices of $25 - 10 = 15$, whereas system 2's value-added in domestic prices is $100 - 60 = 40$. The cost of domestic factors is subsidized in system 1 and taxed in system 2. With the initial policy, system 1 technology is preferred by producers, because it offers higher private profitability than system 2 does (+ 5 versus − 20). Because of its negative private profitability, system 2 does not operate. After the policy change, system 1 does not operate, because private profit is negative, whereas system 2 generates $100 - 60 = 40$ units of foreign exchange. By choosing a policy that encourages operation of the system with the highest social profitability, the country increases direct foreign-exchange generation by $40 - 15 = 25$. This example illustrates that both relative domestic factor costs and relative tradable-input costs are essential elements in the estimation of foreign-exchange effects. A shift in policy toward greater efficiency increases the availability of foreign exchange. Therefore, a foreign-exchange balance, if pursued correctly, need not be a nonefficiency objective.

Concluding Comments

For most developing countries, a critical macro-micro linkage in policy analysis is the taxing effect of an overvalued exchange rate and the ability of privileged producers of tradables to obtain insulating trade restrictions. Very few groups in a developing country benefit in the short run from deflationary macroeconomic policy, and only those ultimately taxed by overvaluation are short-run gainers from devaluation. Because long-run benefits are hard to sell politically in any society, macroeconomic reform typically occurs only when a country has exhausted all delaying options. Hence, it is much easier to steer a developing economy off its best macroeconomic policy path to long-run development than it is to make the painful corrections usually needed to return to that path.

This unfortunate reality causes special difficulties for agriculture and rural-based small industry in countries with distorted macroeconomic policy. The tradable outputs of rural producers are taxed by the overvalued exchange rate, but rural entrepreneurs generally do not have the political clout to receive quantitative protection to negate that macroeconomic disincentive. Therefore rural interests in distorted economies feel the taxing effects of macroeconomic policy, whereas urban industrialists receive insulating protection through effective quantitative restrictions. The few developing countries that successfully protect their agricultural producers with import quotas as part of price stabilization programs do not usually experience large and persistent exchange-rate overvaluation; not coincidentally, in those countries, successful commodity policy is accompanied by nondistorting macroeconomic policy.

Bibliographical Note to Chapter 5

An introduction to the impact of macroeconomic policy on food systems in developing countries is provided in C. Peter Timmer, Walter P. Falcon, and

Scott R. Pearson, *Food Policy Analysis* (Baltimore: Johns Hopkins University Press, 1983), pp. 215–59. A detailed discussion of public finance issues, covering the development of public sector budgets and the alternative means to finance the budget, is contained in Richard Goode, *Government Finance in Developing Countries* (Washington, D.C.: Brookings Institution, 1984). Works that focus on revenue generation include Stephen R. Lewis, Jr., *Taxation for Development* (New York: Oxford University Press, 1984), and John F. Due, *Indirect Taxation in Developing Economies* (Baltimore: Johns Hopkins University Press, 1974). Agricultural taxation issues receive detailed discussion in Richard M. Bird, *Taxing Agricultural Land in Developing Countries* (Cambridge: Harvard University Press, 1974).

An introduction to the issues of exchange-rate determination and policy is provided in Anne O. Krueger, *Exchange Rate Determination* (Cambridge: Cambridge University Press, 1983), a paperback volume in the Cambridge Surveys of Economic Literature series. Further exposition at the intermediate textbook level is contained in pts. 4 and 5 of Richard E. Caves and Ronald W. Jones, *World Trade and Payments: An Introduction* (Boston: Little, Brown, 1984), pp. 277–541. Those wishing to delve more deeply into specific aspects of exchange-rate economics are referred to Ronald W. Jones and Peter B. Kenen, eds., *Handbook of International Economics* (Amsterdam: North-Holland, 1984), vol. 2, for extensive surveys and complete listings of citations.

The Australian model of exchange-rate adjustment, which focuses on the distinction between traded and nontraded goods and stresses the need to use the two policies of devaluation and deflation to achieve the two targets of external balance and internal balance, is described concisely in W. M. Corden, *Inflation, Exchange Rates, and the World Economy,* 3d ed. (Chicago: University of Chicago Press, 1986), especially pp. 7–33. That analysis is extended to the role of exchange-rate adjustment in economies with rapidly increasing export earnings in Corden, "Booming Sector and Dutch Disease Economies: Survey and Consolidation," *Oxford Economic Papers* 35 (November 1984): 359–80. This topic is treated from the viewpoint of a developing country with a large agricultural sector in Corden and P. G. Warr, "The Petroleum Boom and Exchange Rate Policy in Indonesia: A Theoretical Analysis," *Economics and Finance in Indonesia* 29 (September 1981): 335–59.

Empirical evidence on the relationship between exchange rates and inflation is analyzed in Ronald I. McKinnon, *Money in International Exchange: The Convertible Currency System* (New York: Oxford University Press, 1979). A seminal article that investigates the links between exchange-rate policy and the agricultural sector is G. Edward Schuh, "The Exchange Rate and U.S. Agriculture," *American Journal of Agricultural Economics* 56 (February 1974): 1–13.

Two theoretical essays by W. M. Corden that lay out the central issues of protection afforded by exchange-rate policy and by commodity-specific policy are "Exchange Rate Protection," pp. 271–87, and "Protection and the Real Exchange Rate," pp. 302–310, in Corden, *Protection, Growth and Trade:*

Essays in International Economics (Oxford: Basil Blackwell, 1985). A useful analysis of the combined impact of exchange-rate and commodity price policies on agricultural production and trade is found in the World Bank's *World Development Report, 1986* (New York: Oxford University Press, 1985), pp. 61–168; the report also has a detailed list of references. Many empirical studies of the relationships between the macroeconomic policies, especially exchange-rate policies, and the economic performance of developing countries have been carried out, notably by research teams led by Bela Balassa, Jagdish Bhagwati, and Anne Krueger. The work done in the 1970s was reviewed and synthesized in Ronald I. McKinnon, "Foreign Trade Regimes and Economic Development: A Review Article," *Journal of International Economics* 9 (August 1979): 429–52.

PART III

SOCIAL VALUATION IN THE POLICY ANALYSIS MATRIX

CHAPTER 6

Social Valuation of Commodities

WORLD PRICES AND TECHNOLOGIES are the backbones of social valuation and efficiency analysis of agricultural systems. The first section of this chapter uses the simple general equilibrium model of production to show how an economy attains its highest levels of output and income by using world prices. In this process of maximization, factor prices are determined, providing the basis for social valuation of new commodity systems. If a new commodity system is unable to pay the social costs of domestic factor inputs under world prices, national income could be increased by leaving domestic factors in some other commodity system.

Because world prices are quoted in foreign currency, a foreign-exchange rate is needed to convert world prices from foreign to domestic currency. Typically, entries in the PAM matrix are presented in domestic currency, because data on revenues and costs in private prices are collected in domestic currency and corrections for market failures and estimates of social prices of factors are easier to make in local currency than in foreign currency. Estimation of the PAM in terms of foreign currency would provide no shortcuts, because a foreign-exchange rate would be needed to convert all domestic currency values to foreign currency equivalents. The issue then becomes whether the official exchange rate—if fixed by the government and administered by the central bank—is an appropriate choice for conversion. The second section of this chapter considers the circumstances in which other exchange-rate measures might be used for social price determination. The final section considers the treatment of nontradable commodities. Because these commodities lack world markets (by definition), social valuation of nontradables is based on calculations of social costs of production.

Social Prices in a General Equilibrium Model

The two-good, two-factor model of international trade provides the simplest framework in which to establish a basis for social price determination. Figure 6.1 illustrates an economy capable of producing two goods, Q_1 and Q_2, with fixed supplies of two domestic-factor inputs, labor and capital. Both inputs are necessary in the production of either good, and both may be shifted from one industry to the other. The production of both goods takes place with constant returns to scale; for example, doubling the amount of each input in an industry results in doubling the amount of output. These two assumptions ensure that the production possibilities curve is at least as far from the origin as the straight line EZF. Putting all labor and capital in industry Q_1 results in an output of E, whereas placing all inputs in industry Q_2 results in an output of F. Using half of each input in each industry results in the output combination at point Z, which represents half the maximum outputs of Q_1 and Q_2.

World Prices and Maximum Consumption

The economy is likely to be capable of better performance than that indicated by line EZF. If inputs have different productivities in the two industries, total output can be increased by allowing different input allocations between the commodities. When inputs are reallocated between the two industries, the output of Q_1 can be maintained and the output of Q_2 can be increased, resulting in an output combination represented by Y. The assumption of diminishing marginal returns to each input means that this set of maximum production points will have a shape that is concave with respect to the origin. When the economy is completely specialized in production of the first commodity (point E), labor and capital resources can be withdrawn with relatively little impact on the output of good 1 and a relatively large impact on the output of good 2. But with diminishing marginal productivities, the incremental tradeoffs become progressively less attractive. Successively larger amounts of good 1 must be sacrificed to attain a one-unit increase in the output of good 2. The maximum production possibilities frontier is represented by the curve EBYF.

Movements along the production possibilities curve express the opportunity cost of one good's production in terms of the other good. The curve can be interpreted also as the consumption possibilities for an economy that is entirely self-sufficient. But the introduction of interna-

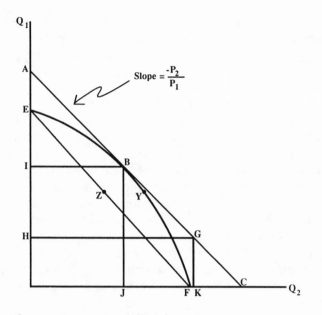

Figure 6.1. The consumption possibilities frontier

tional trading opportunities expands the consumption possibilities set beyond EBYF. If the country is too small to influence commodity prices, the trading opportunities of the economy can be represented by straight lines that intersect the relevant production point. One such line is ABGC. All points along this line represent combinations of goods 1 and 2 that have an equal value in international markets. Therefore, $(\Delta Q_1 \times P_1) = -(\Delta Q_2 \times P_2)$. Rearranging the terms gives $(\Delta Q_1/\Delta Q_2) = (-P_2/P_1)$. This result shows that the slope of the trading opportunities line, $\Delta Q_1/\Delta Q_2$, can be expressed also as the negative ratio of world prices, $-P_2/P_1$. The choice of which good to import and which to export then depends on domestic consumer preferences. Consumption at point G implies exports of Q_1 equal to HI and imports of Q_2 equal to JK. The choice of a point along segment AB would imply imports of Q_1 and exports of Q_2.

The line ABGC is the maximum consumption possibilities frontier for the economy. No other trade opportunity line (of slope $-P_2/P_1$) would include a point on the production possibilities curve and still allow such large amounts of the two commodities to be consumed. Because production income equals expenditure on consumption, this maximum can be measured by evaluation of either consumption choices at world prices or production choices at world prices. The evaluation of production involves a unique point (B). But the economy can choose to consume at any point on ABGC because these points are all of equal value.

Factor Prices

World prices are the social prices for tradable commodities because their use allows the economy to reach the maximum consumption possibilities frontier. The remaining social prices needed for the simple model are the rental rate (interest plus depreciation) for capital, r, and the wage rate for labor, w. Under the assumption that factor supplies are fixed, these input prices are determined by the prices for outputs and production technology. The assumption of constant returns to scale means that so long as both outputs are produced, knowledge of the particular amounts of Q_1 and Q_2 is not necessary for the estimation of social input prices. Only world prices and technology matter.

If competition for the services of domestic factors eliminates excess profits, total costs and total revenues can be expressed as an equality, as in equation 1:

$$wL_1 + rK_1 = P_1Q_1$$
$$wL_2 + rK_2 = P_2Q_2 \tag{1}$$

where L and K represent quantities of labor and capital. Division by Q_1 and Q_2 yields equation 2:

$$w\frac{L_1}{Q_1} + r\frac{K_1}{Q_1} = P_1$$
$$w\frac{L_2}{Q_2} + r\frac{K_2}{Q_2} = P_2 \tag{2}$$

The ratios L/Q and K/Q represent the use of the inputs per unit of output. Equation 3 reformulates equation 2, using l_i and k_i as input-output coefficients:

$$wl_1 + rk_1 = P_1$$
$$wl_2 + rk_2 = P_2 \tag{3}$$

Equation 3 is termed the zero-profit condition. It is meant to represent a state that the economy would occupy if prices and technologies did not change over time. In reality, industries rarely demonstrate zero profits. Instead, output prices and technologies change frequently, with the result that total costs are usually greater than or less than total revenues. The use of the zero-profit condition as a basis for social price determination thus measures incentives under long-run equilibrium, assuming a continuation of the conditions prevailing during a particular time period.

Rearrangement of the zero-profit condition (equation 3) shows how the prices of inputs are dependent on the output prices and the input-output coefficients:

$$w = \frac{P_1 k_2 - P_2 k_1}{l_1 k_2 - l_2 k_1}$$

$$r = \frac{P_2 l_1 - P_1 l_2}{l_1 k_2 - l_2 k_1} \tag{4}$$

These factor costs represent the social opportunity costs of factors used by a new commodity system, Q_3. If production of Q_3 cannot profitably compensate domestic factors at prices w and r, national income cannot be increased from the introduction of Q_3. The economy would be better off, in the sense of maximum consumption possibilities, by remaining with the production of goods Q_1 and Q_2. The calculation of PAM usually treats the system under study as a new commodity system, so that social opportunity costs of factors are determined by the other commodities in the economy. Social profit then represents the net contribution of the commodity system to national income.

Generalization of the Simple Model

The input-output equations of the simple model can be expanded to provide a model of any desired degree of detail. Equation 5 illustrates this general model of the economy:

$$w_1 Z_{11} + w_2 Z_{21} + w_3 Z_{31} + \ldots + w_m Z_{m1} = P_1 Q_1$$
$$w_1 Z_{12} + w_2 Z_{22} + w_3 Z_{32} + \ldots + w_m Z_{m2} = P_2 Q_2$$
$$\vdots \qquad \vdots \qquad \vdots \qquad\qquad \vdots \qquad \vdots$$
$$\vdots \qquad \vdots \qquad \vdots \qquad\qquad \vdots \qquad \vdots$$
$$w_1 Z_{1n} + w_2 Z_{2n} + w_3 Z_{3n} + \ldots + w_m Z_{mn} = P_n Q_n \tag{5}$$

where
w_i = Price of input i
p_j = Price of output j
Q_j = Quantity of output j
Z_{ij} = Quantity of input i used in production of output j

Dividing through the jth equation by Q_j yields the analog of equation 3:

$$w_1a_{11} + w_2a_{21} + w_3a_{31} + \ldots w_m a_{m1} = P_1$$
$$w_1a_{12} + w_2a_{22} + w_3a_{32} + \ldots w_m a_{m2} = P_2$$

$$w_1a_{1n} + w_2a_{2n} + w_3a_{3n} + \ldots w_m a_{mn} = P_n \tag{6}$$

Some of the inputs used in the production of goods will also be tradable goods; fertilizers, seeds, and agricultural chemicals are examples of commodities available on world markets. If Z_1 through Z_4 are assumed to be domestic factor inputs and Z_5 through Z_m are assumed to be tradable inputs, the equations can be rewritten so that domestic factor inputs are segregated from tradable outputs and inputs, as shown in equation 7:

$$w_1a_{11} + w_2a_{21} + w_3a_{31} + w_4a_{41} = P_1 - P_5a_{51} - \ldots - P_m a_{m1}$$

$$w_1a_{1n} + w_2a_{2n} + w_3a_{3n} + w_4a_{4n} = P_n - P_5a_{5n} - \ldots - P_m a_{mn} \tag{7}$$

where $w_5, w_6, \ldots, w_m$ are replaced by world prices $P_5, P_6, \ldots, P_m$ because these inputs are also tradable outputs.

The right-hand side of the equation now represents value added rather than output price. In matrix form, equation 7 can be written as follows:

$$[w_1 \ w_2 \ w_3 \ w_4] \begin{bmatrix} a_{11} \ a_{12} \ldots a_{1n} \\ a_{21} \ a_{22} \ldots a_{2n} \\ a_{31} \ a_{32} \ldots a_{3n} \\ a_{41} \ a_{42} \ldots a_{4n} \end{bmatrix} = [VA_1 \ VA_2 \ldots VA_n] \tag{8}$$

Equation 8 is another form of the zero-profit condition. The prices of domestic factors, times the relevant input-output coefficients, exactly equals each VA_j, the value-added in production. Given world prices and input-output coefficients, the domestic factor prices can be determined in a manner analogous to that used in the simple model. But in order for the system of equations to generate a solution, the matrix of input-output coefficients must be square. The number of commodities must equal the number of factors. This requirement might appear to compromise the generality of the approach, as most economies are unlikely to attain such an equality. In equation 8, for example, only four commodities are needed to determine the factor prices.

Counting numbers of inputs and outputs is a difficult and ultimately arbitrary exercise. The number of domestic factor inputs can be made almost infinitely large if one recognizes different types of labor, capital, and land. The number of commodities can also be expanded if sector output is divided into commodity groups, specific commodities, or brand names and qualities of a particular commodity. But social factor prices can still be determined, as long as the number of goods produced in the economy exceeds the number of domestic factors engaged in their production. The presence of more goods than factors thus implies redundant information for the determination of social factor prices. Calculation of the social factor prices involves a search for the core set of production activities that will result in the maximum income in the economy. Any activity that pays domestic factors less than their values attained under the maximum income for the economy will be eliminated by competition for the use of factor inputs. The circumstance of more commodities than factors appears reasonable for the majority of empirical circumstances.

Exchange Rates and Social Valuation

In many economies, exchange rates are controlled or influenced by government policies and thus may bear no resemblance to the rates that would prevail under social pricing conditions. A distorted exchange rate can therefore affect the domestic currency price of tradable commodities. Figure 6.2 illustrates the short-run and long-run effects of exchange-rate change on a tradable-commodity market. The initial conditions are represented by supply curve S and domestic price P_D. In these circumstances, domestic production (Q_P) exceeds domestic consumption (Q_C), and ($Q_P - Q_C$) of the commodity is exported. The domestic price is determined by the world price, measured in foreign currency (P_W) times the exchange rate (e).

This exchange rate is assumed to be distorted (for example, by a pervasive government budget deficit) so that it cannot be sustained in the long run. A depreciation in the domestic currency is needed, raising the exchange rate to e'. As a result, the domestic currency price increases to P'_D. At the same time, the supply curve is affected, since the costs of

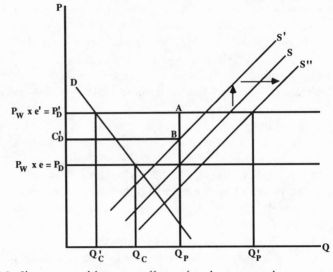

Figure 6.2. Short-run and long-run effects of exchange-rate changes on tradable commodities

tradable-commodity inputs increase. Because these inputs do not account for all production costs, the proportional upward shift in the supply curve will not be as large as the proportional change in output price. In the figure, the supply curve shifts upward to S'. In the short run, prices of tradable outputs and inputs increase, but quantity of output does not; commodity producers thus earn excess profits. The output price is P'_D, marginal costs are C'_D, and excess profits are $(P'_D - C'_D)Q_P$, or $P'_D ABC'_D$.

In response to the excess profits, producers will seek to expand output and thus increase their use of tradable inputs and domestic factors. But producers in all other tradable-commodity markets will be trying the same thing. Because the exchange-rate change creates excess profits in all importable and exportable markets, bidding for the services of domestic factors will be widespread. Prices of domestic factor inputs will rise and will continue to increase until competition for factor services eliminates excess profits. Thus, in the long run, foreign-exchange rates will affect all input prices to the same extent as output prices.

This effect is shown by the formulas for factor price calculations. Private market factor prices are represented by a superscript P and private market commodity prices are represented by a superscript D:

$$w^P = \frac{P_1^D k_2 - P_2^D k_1}{l_1 k_2 - l_2 k_1}, r^P = \frac{P_2^D l_1 - P_1^D l_2}{l_1 k_2 - l_2 k_1}$$

Social factor prices are shown with a superscript S and social commodity prices have a superscript W:

$$w^S = \frac{P_1^W ek_2 - P_2^W ek_1}{l_1 k_2 - l_2 k_1}, \, r^S = \frac{P_2^W el_1 - P_1^W el_2}{l_1 k_2 - l_2 k_1}$$

If divergences are absent, domestic commodity prices equal world prices ($P^P = P^W$), and domestic and social factor prices are equal. If domestic prices are increased above world prices by uniform exchange-rate depreciation, the factor price equation in the private market for labor can be rewritten as

$$w^P = \frac{(P_1^W)e(1 + t)k_2 - (P_2^W)e(1 + t)k_1}{l_1 k_2 - l_2 k_1}$$

Factoring out the $1 + t$ gives

$$\frac{w^P}{(1 + t)} = \frac{P_1^W ek_2 - P_2^W ek_1}{l_1 k_2 - l_2 k_1} = w^S$$

Similar results follow for the price of capital, showing that both social factor prices are derived from private market prices by adjustment for the magnitude of distortion in the exchange rate. If output prices are higher than world prices because of an undervalued exchange rate, so are factor prices. If output prices are lower than world prices because of an overvalued exchange rate, so too are factor prices. Identical results are obtained for the more general model (equation 8).

As domestic factor prices increase, the supply curve in Figure 6.2 will tend to shift upward (from S') toward an intersection with point A. But this point cannot represent the new market equilibrium. Because the increase in domestic factor prices coincides with (and depends on) increased employment of factors, the market supply must shift outward, so that production exceeds Q_P. In the figure, equilibrium is represented by an outward shift to S'' and production expands to Q_P'. Exports are likely to increase as well, although the amount depends on exchange-rate-induced shifts in the domestic demand curve (not shown). Unless domestic factors are unemployed, the inputs necessary to permit the expansion of tradables production must come from the nontradables sector.

In the evaluation of tradable-commodity systems, the calculation of social prices for foreign exchange may represent a needless complication. If the government distorts the exchange rate for the economy, in the long run this rate will influence all tradable-output, tradable-input, and domestic factor prices in equal proportion. Social profitability of the system will differ from private profitability by the same proportional factor. Because exchange-rate adjustments cannot alter the sign of profitability or the profitability rankings of different tradable-commodity systems, exchange-rate distortions are often ignored in social price calculations.

In some instances, however, policy analysts will be concerned with the magnitude of exchange-rate change. First, analysts may want to explain the causes of divergences between private and social prices. As chapter 5 showed, commodity policy and macroeconomic policy are often intertwined. Second, the rate of adjustment of costs of the various categories of inputs is likely to vary. Tradable-input prices are affected quickly, as soon as the domestic prices of imports or exports are altered by the new exchange rate. But the prices of labor, capital, and land adjust more gradually, only after the impacts of increased factor demands from the tradable-goods sector are felt. In the short run, therefore, exchange-rate changes increase profitability in tradable-commodity systems, and PAM analysts interested in short-run incentives might choose to adjust only the social values of tradable commodities for changes in the exchange rate.

Analysts may also be interested to show the change in real income of domestic factors that results from the return to an equilibrium exchange rate. If factor prices ultimately change in the same proportion as the exchange rate, the owners of factor services are potentially better off. But the exchange rate affects prices of tradable commodities as well, and the cost of the consumption basket of each factor increases also. The prices of nontradable commodities are not directly affected, however. For domestic factor owners who consume some nontradable goods, the increase in the cost of the total consumption basket will be less than the increase in factor prices. As a result, real factor prices will be higher and the real income of factor owners will rise after the exchange-rate change, although by a smaller proportion than the change in the exchange rate.

The introduction into the PAM of short-run impacts of exchange-rate changes is straightforward. The social values of tradable outputs and tradable inputs (E and F) are multiplied by the ratio of the equilibrium exchange ratio (e') to the existing exchange rate (e). However, estimation of e' is very demanding of empirical information. Figure 6.2 shows that the contribution of tradable commodities to an improvement in the foreign-exchange balance depends on the slopes and shifts of the supply and demand curves: the own-price elasticity of supply, the cross-price elasticities of supply (especially with respect to nontradables), the own- and cross-price elasticities of demand, and the income elasticities of demand. In Figure 6.2, these shifts cause exports to increase from $Q_P - Q_C$ to $Q'_P - Q'_C$. With aggregation of these effects across commodities, exchange-rate changes can be associated with changes in the aggregate balance of foreign exchange, thus identifying an equilibrium. Because of the substantial information requirements, however, approximations of equilibrium exchange rates will usually be uncertain. Directions of change can be understood with confidence, but magnitudes are likely to be elusive.

Multiple Exchange-Rate Regimes

More complicated adjustments to social commodity prices are needed in the presence of multiple exchange rates. The government might establish a set of exchange rates that differ according to commodity. This result can be achieved directly with a multiple exchange-rate regime or indirectly by placement of quantitative trade restrictions on certain importables or exportables. Quotas allow the protected commodities to be traded at effective exchange rates different from the official rate used for unprotected or tariff-protected goods. Alternatively, parallel markets for foreign exchange might operate along with the official government market. In this circumstance, the effective exchange rates for domestic factor prices may differ from the particular exchange rates used for the tradable commodities of a commodity system.

In the simple general equilibrium model, for example, the wage rate represents a weighted average of the different exchange rates:

$$w = \frac{(e_1 P_1^W)k_2 - (e_2 P_2^W)k_1}{l_1 k_2 - l_2 k_1}$$

where e_1 and e_2 represent the exchange rates applicable for two commodities. The average effective exchange rate can be represented by e_3, where

$$e_3(P_1^W k_2 - P_2^W k_1) = e_1 P_1^W k_2 - e_2 P_2^W k_1.$$

In larger, more realistic models, each exchange rate appears in positive terms and negative terms of the factor price solution. For example, the three-good, three-factor model can be described in equation system 9:

$$P_1^D = wl_1 + rk_1 + zt_1$$
$$P_2^D = wl_2 + rk_2 + zt_2$$
$$P_2^D = wl_2 + rk_2 + zt_2 \tag{9}$$

where z is the unit price of the third factor and t is the input-output coefficient. If each commodity is traded at a different exchange rate, the average exchange rate for labor cost is e_4, where

$$e_4 = \frac{(e_1 P_1^W k_2 t_3 + e_2 P_2^W k_3 t_1 + e_3 P_3^W k_1 t_2 - e_1 P_1^W k_3 t_2 - e_2 P_2^W k_1 t_3 - e_3 P_3^W k_2 t_1)}{(P_1^W k_2 t_3 + P_2^W k_3 t_1 + P_3^W k_1 t_2 - P_1^W k_3 t_2 - P_2^W k_1 t_3 - P_3^W k_2 t_1)}$$

This expression may be rewritten as

$$e_4 = e_1 \frac{P_1^W(k_2 t_3 - k_3 t_2)}{Y} + e_2 \frac{P_2^W(k_3 t_1 - k_1 t_3)}{Y} + e_3 \frac{P_3^W(k_1 t_2 - k_2 t_1)}{Y}$$

where

$$Y = P_1^W(k_2 t_3 - k_3 t_2) + P_2^W(k_3 t_1 - k_1 t_3) + P_3^W(k_1 t_2 - k_2 t_1)$$

This expression shows that the value of e_4 depends on the various world prices and input-output coefficients. If each of the weights on e_1, e_2, and e_3 is less than 1 (because the value of Y exceeds the value of each of its individual terms), e_4 will lie somewhere within the observed range of multiple exchange rates.

Assuming conversion ratios lie within the observed range of multiple rates, approximate values can be based on the relative prominence of the different exchange rates. If one particular exchange rate dominates transactions, most of the terms in the formulas would have a single exchange rate and the other terms would not be sufficiently numerous to generate an average value very different from the dominant value. If most commodities are traded at rate e_1 and only a few commodities are traded at a lower rate—e_2, for example—the average rate selected, e_3, would be slightly less than e_1. Therefore, for domestic currency values of tradable commodities that have been exchanged at rate e_1, social values would be reduced by the value of e_3/e_1. The social value of tradable commodities that have been exchanged at rate e_2 would be increased by e_3/e_2.

Nontradable Goods and Social Valuation

The general model solution for social factor prices is based on the assumption that all commodities have world market prices. But in almost all economies, a large class of goods is not traded on interna-

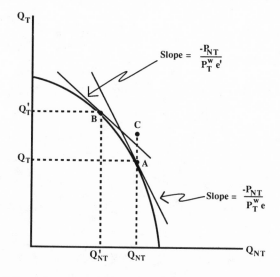

Figure 6.3. Relative prices of tradable and nontradable goods

tional markets. Nontradable goods such as electricity and water have high international transport costs. Nontradable services such as marketing activities and legal services are impossible to supply from foreign sources for logistical reasons. For the valuation of social factor prices, these commodities are considered redundant goods. Because nontradable outputs do not face international competition, the domestic market price will alter to any degree necessary to cover costs. The factor prices paid by the tradable-goods sector must also be paid to factors used by the nontradable-goods sector.

However, social prices for nontradables remain complicated by the impacts of domestic demand shifts. For tradable commodities, domestic demand shifts are usually ignored because they affect only the magnitude of imports or exports, not the world price. For nontradable goods, demand shifts from changes in incomes or relative prices—in turn caused by the elimination of commodity market or exchange-rate divergences—will directly affect social output prices and production costs.

Figures 6.3 and 6.4 demonstrate this result for the case of exchange-rate depreciation. In Figure 6.3, production is initially represented by point A; the initial exchange rate is e. This production point is not sustainable because the exchange rate is overvalued. Consumption of nontradables equals production, but consumption of tradables (point C) exceeds the production. To resolve this disequilibrium, the exchange rate depreciates to e', raising the domestic currency price of tradables. Production shifts to point B. If deflationary policy forces total expendi-

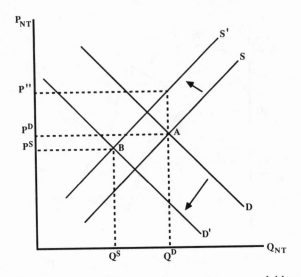

Figure 6.4. Exchange rates and real-price adjustments in nontradable goods

ture to decline at the same time so as to become equal to production income (through reduced government spending, higher taxes, or higher interest rates that reduce consumption), this point represents a sustainable equilibrium for the economy. Tradable-goods production has increased from Q to Q', whereas nontradable-goods production has declined from Q_{NT} to Q'_{NT}.

Figure 6.4 provides a partial equilibrium interpretation of the effect of an exchange-rate change on a nontradable-commodity market. Initial equilibrium at point A is accompanied by price P^D and output Q^D. A change in the exchange rate from a distorted to a social value will cause tradable-input prices to rise, shifting the supply curve upward in the short run. In the long run, the supply curve must shift upward even further, reflecting increases in domestic factor costs, and leftward, because of the exit of resources from the sector. The new long-run supply curve becomes S'. Because the relative price of nontradables must decline in the new equilibrium, the reductions in total expenditure must shift the demand curve to more than offset the shift in the supply curve. The new equilibrium (point B) results in a lower price, P^S, and a smaller quantity, Q^S, than in the initial situation. The new price represents the social value of the nontraded output. Because of the effects of demand shifts, evaluation of an observed system in terms of social input prices may not be sufficient. In Figure 6.4, such a procedure results in a social price for output of P'', above the true social value of P^S.

Concluding Comments

The social prices for goods and factors are associated with the max-imization of aggregate income of an economy operating with competi-tive markets for outputs and inputs. For tradable commodities, com-modity-specific policy and market failures are the principal source of difference between private and social commodity prices. Unless the analyst requires results concerning short-run conditions, the econo-mywide distortions of exchange rates will have a uniform impact on tradable-commodity systems. Some complications result from complex exchange management policies, but they are manageable in most cir-cumstances. For social pricing of nontradable commodities, however, the estimation problems are far more difficult. For these commodities, explicit consideration of domestic demand and exchange-rate policy become essential elements of the social pricing exercise.

Bibliographical Note to Chapter 6

The principal results of the simple model of international trade—that world prices are associated with maximum consumption and that output prices (with technology) determine input prices—are discussed in Paul Samuelson, "The Gains from International Trade Once Again," *Economic Journal* 72 (December 1962): 820–29; and Ronald W. Jones, "The Structure of Simple General Equilibrium Models," *Journal of Political Economy* 73 (December 1965): 557–72. Generalization of the two-good, two-factor model has received sub-stantial attention from international trade theorists. Introductions to and sur-veys of these results are provided in Ronald W. Jones, "Two-Ness in Trade Theory: Costs and Benefits," *Special Papers in International Economics*, no. 12 (Princeton, N.J.: Department of Economics, Princeton University, 1977); and Wilfred Ethier, "Higher Dimensional Issues in Trade Theory," chap. 3 in *Handbook of International Economics,* ed. Ronald Jones and Peter Kenen (Amsterdam: North-Holland, 1984).

Generalization of the social pricing model requires identification of the optimal set of outputs associated with maximum national income. Arbitrary selection of *n* industries (when there are *n* factors) creates a large number of potential sets of shadow factor prices: Trent Bertrand, "Shadow Pricing in Distorted Economies," *American Economic Review* 69 (December 1979): 902–14. Many economies have industries that exist only because of the support of policy, and inclusion of these industries would generate shadow prices that support inefficient industries. The empirical difficulty is to locate the set of industries that maximizes total consumption. In this maximization process, the quantity produced of each output may not be determinate: Jaroslev Vanek and

Trent Bertrand, "Trade and Factor Prices in a Multi-Commodity World," in *Trade, Balance of Payments and Growth*, ed. Jagdish Bhagwati et al. (Amsterdam: North-Holland, 1971). However, because factor prices depend only on output prices, variations in total production (and in factor endowments) will not affect the factor prices. These arguments and their qualifications are discussed in Jagdish Bhagwati and H. Wan, Jr., "The 'Stationarity' of Shadow Prices of Factors in Project Evaluation, with and without Distortions," *American Economic Review* 69 (June 1979): 261–73.

A second issue, the unique correspondence between output prices and factor prices, has been a principal concern of the literature on Samuelson's factor-price equalization theorem. Uniqueness requires restrictions (known as Gale-Nikaido conditions) on the mathematical properties of the matrix of input-output coefficients. Two survey articles review this literature: Ethier, "Higher Dimensional Issues"; and Ronald W. Jones and J. Peter Neary, "The Positive Theory of International Trade," chap. 1 in Jones and Kenen, *Handbook of International Economics*.

All empirical estimations of general equilibrium models inevitably require aggregation of commodities into groups, and the literature on aggregation and commodity separability discusses the strong assumptions involved in this exercise. Angus Deaton and John Muellbauer, *Economics and Consumer Behavior* (New York: Cambridge University Press, 1980), chap. 5, focuses on consumer demand, but necessary production relationships are analogous. A more detailed examination is contained in Charles Blackorby, D. Primont, and R. R. Russell, *Duality, Separability and Functional Structure* (New York: American Elsevier, 1978).

Analysis of exchange rates and nontradables began to receive increased attention in the 1970s, with the termination of fixed-exchange-rate regimes in most developed countries. Many of the references listed in chapter 5 are relevant here, particularly W. M. Corden, *Inflation, Exchange Rates and the World Economy*, 3d ed. (Chicago: University of Chicago Press, 1986). Another wide-ranging discussion of exchange-rate issues is contained in Rudiger Dornbusch, *Open Economy Macroeconomics* (New York: Basic Books, 1980). Works that emphasize the linkages between exchange rates and nontradable goods are Rudiger Dornbusch, "Devaluation, Money and Nontraded Goods," *American Economic Review* 73 (December 1973): 871–80; Dornbusch, "Real and Monetary Aspects of the Effects of Exchange Rate Changes," in *National Monetary Policies and the International Financial System*, ed R. Z. Aliber (Chicago: University of Chicago Press, 1974), pp. 64–81; and Michael Bruno, "The Two-Sector Open Economy and the Real Exchange Rate," *American Economic Review* 66 (September 1976): 566–77.

The use of world prices as the basis for social prices has also been important in cost-benefit analysis. One of the earliest works is I. M. D. Little and J. A. Mirrlees, *Project Appraisal and Planning for Developing Countries* (New York: Basic Books, 1974). A summary of this approach is contained in V. Joshi,

"The Rationale and Relevance of the Little-Mirrless Criterion," *Oxford Bulletin of Economics and Statistics* 34 (February 1972): 3–32. Numerous other works have elaborated on the use of world prices as cost-benefit tools in the presence of a wide variety of domestic divergences and public sector policy constraints. Even in the second-best scenarios, world prices usually emerge as appropriate social prices. Two of these works are Partha Dasgupta and Joseph E. Stiglitz, "Benefit-Cost Analysis and Trade Policies," *Journal of Political Economy* 82 (January/February 1974): 1–33; and Peter G. Warr, "On the Shadow Pricing of Traded Commodities," *Journal of Political Economy* 85 (August 1977): 865–72.

CHAPTER 7

Social Valuation of Factors

THIS CHAPTER discusses the determination of social factor prices in the general equilibrium model. Lack of information often makes the empirical estimation of general equilibrium models difficult or impossible. However, the PAM identity—divergences equal the difference between private and social values—suggests a less direct approach to estimation. Beginning with observed (private) factor prices, one derives social prices by adjusting these prices for the impacts of divergences. Factor prices are affected directly by policy-imposed distortions in factor markets (such as rent controls, interest-rate regulations, and minimum wages) and by factor market imperfections (such as monopsony or monopoly power). Observed factor prices can also be affected indirectly by commodity market divergences and macroeconomic distortions. Furthermore, the patterns of input use can change as producers adjust output levels to social output prices and input combinations to social input prices. If output price response and input substitution occur in a large number of industries, aggregate demand for each factor input can shift, altering factor prices.

Price Determination in Factor Markets

Social valuation of domestic factors of production differs from that of tradable commodities because factors are assumed to be immobile across national borders. In principle, international migration opportunities could dominate domestic factor markets, and social factor prices would be determined externally, like tradable-commodity prices. In a country with a completely liberalized capital market, for example, no restrictions are made on domestic investment behavior. The investor is free to choose between domestic and foreign investment opportunities. Because the world capital market is much larger than the domestic market, the rate of return to investment can be considered exogenous to the domestic economy. Domestic distortions in capital markets can be ignored. Such distortions affect only the magnitude of capital imports or exports, not the social value of capital. Similarly, if the emigration of workers is particularly widespread, the domestic economy must offer wage rates equal to foreign-determined opportunity costs. Again, knowledge of domestic market divergences becomes unnecessary in the calculation of social values.

Among domestic factors, only land is universally immobile. But in most countries, international migration of labor and capital is also heavily constrained. The constraints reflect cultural differences (such as language and religious differences), prohibitive transactions costs of migration, or foreign-imposed limits on market access. Given these constraints, foreign earning opportunities for domestic factors will exert some effects on domestic factor prices, but factor prices can be considered to be determined in domestic markets.

In the simple general equilibrium model, factor supplies are assumed to be fixed. As a result, factor price determination is driven entirely by factor demand. But it is easy to incorporate some supply-side influences into social price selection. One unrealistic aspect of the vertical supply curve is its intercept with the x-axis, implying that factor prices could fall to zero without affecting factor supply. In reality, each market is probably characterized by minimum floors for factor prices, below which none of the factor is supplied. These floors can be related to subsistence income needs, the presence of minimum values of leisure time, or the costs of adjustment of and entry into the market.

Figure 7.1 illustrates the impact of a minimum subsistence wage on the market for unskilled labor. The market supply curve is represented by the right-angled supply curve $w^{sub}AS$ instead of Q^SS. The initial demand curve is represented by D. Aggregate labor demand is Q^D at the minimum reservation wage. The result is unemployment equal to $Q^S - Q^D$. Wages could theoretically fall to the market-clearing level, w^*, but no laborers are willing to be employed at that low, subsubsistence wage. Unemployment can be eliminated only if demand increases, to D', for example. Such increases might result from labor-intensive

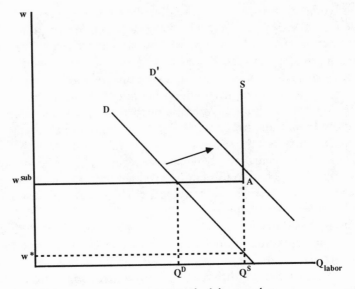

Figure 7.1. Minimum reservation prices in the labor market

technical changes, output expansion, or increased output prices. If labor supplies are increasing over time, such changes are essential if unemployment is to decline.

This case represents "natural" unemployment that cannot be attributed to any distortions or market failures. Consequently, such factors can be omitted from social factor price valuation. The factors that are fully employed create a constraint on total output, and potential national income can increase only if the supply of fully employed factors is increased. In the process, the demand for complementary unemployed factors will increase as well. But the market prices for the surplus factors will be unaffected. The opportunity costs to the economy of these factors are their reservation prices—subsistence wages for labor or the cost of preparing capital and land for productive uses.

Divergences and Social Factor Prices

Under competitive market conditions, the prices of fully employed factors will reflect marginal value products, unless divergences in the factor market are present. If D represents the price-equivalent value of factor market divergences, observed factor prices can be expressed as

$$w^P = P^P(MPP^P) + D, \tag{1}$$

where w = factor price, P = output price, and MPP = marginal physical product $(\Delta Q / \Delta L)$. A superscript P is used to denote that the variable is observed under private market conditions.

The observed values of P^P and MPP^P in equation 1 might also be distorted from their social values. If t is used to represent the divergence in output prices and ΔMPP to represent the divergence in marginal physical products, equation 1 can be rewritten as

$$w^P = (P^W + t)(MPP^W + \Delta MPP^{CMD} + \Delta MPP^{FMD}) + D \tag{2}$$

where a superscript w is used to indicate the value of a variable under world prices for outputs, ΔMPP^{CMD} represents the effect of commodity market distortions on factor productivity, and ΔMPP^{FMD} represents the effect of factor market distortions on factor productivity. Equation 2 can be rearranged to yield

$$w^P = (P^W \times MPP^W) + (t \times MPP^W) + (P^W + t)(\Delta MPP^{CMD})$$
$$+ (P^W + t)(\Delta MPP^{FMD}) + D \tag{3}$$

The social factor price, w^s, is the value of the marginal product measured at world prices; this is the first term on the right-hand side of equation 3, $(P^W * MPP^W)$.

If equation 3 is rearranged and the interaction terms $t \times \Delta MPP$ are ignored, the following equation results:

$$w^P - w^S = (t \times MPP^W) + (P^W \times \Delta MPP^{CMD}) + (P^W \times \Delta MPP^{FMD}) + D \tag{4}$$

Equation 4 shows that the difference between private (observed) factor prices and their social values can be accounted for by two categories of effects. The term $t \times MPP^W + P^W \times \Delta MPP^{CMD}$ represents the impact of commodity market divergences on factor prices. The term $P^W \times \Delta MPP^{FMD} + D$ describes the effects of factor market divergences; these divergences influence factor prices directly and perhaps indirectly as well, through their effects on input productivity. Fixing high factor prices by legislative decree, for example, encourages producers to reduce use of the input and results in artificially high marginal productivities.

The disaggregated view of factor price divergence presented in equation 4 provides an organizational framework for the evaluation of shadow prices. One begins with observable private market wages, rates of return to capital, and land rents, then identifies various divergences, and finally makes judgments on their quantitative significance. Ideal conditions for the calculation of social factor prices arise when empirical estimates of factor demand and supply curves are available for each industry in the economy. Private market prices and quantities can be combined with information about divergences to determine the shifts or movements along the demand and supply curves in each factor market. When such information is not available or when estimates are considered unreliable, less precise estimates of social values must be formulated.

The following sections consider the two categories of divergences in equation 4: factor market divergences and commodity market divergences. The discussion focuses on computational issues. Subsequent sections consider two groups of indirect influence on the factor prices. Macroeconomic distortions affect factor prices through exchange-rate induced influences on commodity prices or through direct impacts on the price of domestic capital resources. Input substitution effects represent responses to changes in relative factor prices and account for part of the observed changes in marginal physical products described in equation 4.

Factor Market Divergences

Adjustments for the impact of factor market distortions are easiest for proportional taxes or subsidies. In that case, the analyst need only decide whether the taxes or subsidies have been passed on to the factor. For example, social security taxes are often levied on employers in an attempt to increase remuneration to labor. If laborers compete for employment, however, money wages would fall by the full amount of the tax. Assuming that workers eventually receive the value of the social security tax, total compensation remains constant. Only the temporal pattern of wage receipts has changed; labor forgoes some current income in favor of increased payments during retirement or illness-induced absence from work.

When taxes have been applied to all sectors of an economy, decisions on the treatment of employer taxes are based on the consideration of employment levels in the presence of the factor tax. Full employment implies that the tax reduces money wages. When the tax is applied only to certain sectors of the economy (industry but not agriculture, for example), full employment is not sufficient evidence to disregard the tax as a distortion; policy may have caused excess labor to migrate from the taxed sector to the untaxed sector, resulting in full employment. Such movements should cause differences between sectoral factor prices. Comparisons of tax-inclusive wage rates with wage rates in sectors of the economy that do not pay factor taxes indicate whether the taxes should be treated as a divergence (if the two prices are unequal) or ignored (if the two prices are equal).

Adjustments for regulations that fix the absolute level of prices in the factor markets are more difficult. Private market values are sometimes available from parallel markets, and wage rates and land rental rates can often be compared to official prices in order to determine whether the regulations are enforced. But unless parallel markets are large, their prices will not be close to social (nonregulated) values. For example, interest rates in the parallel market can be difficult to relate to the rate of return on investment, particularly if capital markets are subject to fragmentation or other divergences besides rationing.

Market failures are the final category of factor market divergences. These failures are often identified by regional comparisons of prices for the factors. If factors are mobile between regions, integration of the factor market is possible and factor prices in one area may be linked to factor prices in another area. But in general, more direct confirmation of market failures is needed, because competitive factor market circumstances may also explain factor price differentials.

If the costs of migration—transportation and moving costs as well as psychic costs—from one area to another are positive, even perfectly competitive factor prices can differ. Consequently, social prices for the factor need not be equal in all regions. Figure 7.2 illustrates this point. The labor market in region 1 is compared to that in region 2; the wage rate is represented by w^*. The costs of migrating from region 1 to region 2 are $w^* - w^L$; the costs of migrating to region 1 are $w^U - w^*$. Initially, local demand and supply in region 1 are assumed to be in equilibrium at the wage rate w^*, which is equal to the wage rate in region 2. But this equality need not be maintained. Following an increase in local demand in region 1, caused by higher prices of region 1 outputs, demand expands to D' and the wage rate rises to w^1. This rise has no effect on the wage rate in region 2, because the costs of migrating to region 1 are larger than the regional wage difference. Migration will begin only if demand causes the wage rate to rise above w^U. At that point, regional wages will begin to move together. Analogous results follow for the case when demand shifts backward and the local wage rate declines. Regional wage rates become linked only when the wage

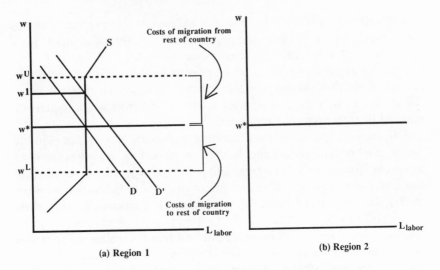

Figure 7.2. Costs of migration and regional variations in the social wage rate

rate falls below w^L, the wage rate that motivates out-migration from region 1.

Another influence on competitive factor price differentials is the duration of employment alternatives. Labor demands and wages may vary seasonally, and employment alternatives may comprise long periods of low-wage work in one region versus shorter periods at higher wages in another region. Factor prices then differ across regions, but market failures need not be present. Total compensation over the course of a production cycle is not different enough to induce migration.

Identification of capital market failures requires a different definition of migration costs and market segmentation. The physical costs of transferring financial capital among regions are extremely small and would not generate any substantial difference in regional rates of return. A single social rate of return, therefore, would be representative for all regions. But social (and private) rates of return can differ more substantially if investment risks vary among potential borrowers. Then the relevant market boundaries are not geographical regions but types of borrowers and commodities. If small borrowers or particular commodities have relatively high probabilities of financial default, rates of return in those sectors must be higher to account for the increased costs of lending. Similarly, some portion of the transaction costs of lending and borrowing are independent of the amount of loan; in percentage terms, small investments must earn higher rates of return than large investments. These differential risks and transaction costs remain even

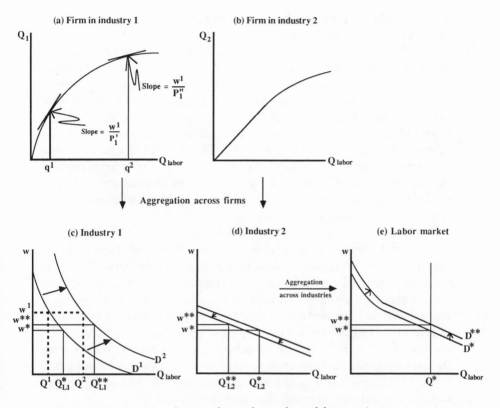

Figure 7.3. Output price changes, factor demands, and factor prices

if capital markets are integrated. Social rates of return need not be equal for all commodity systems.

Commodity Market Divergences

Figure 7.3 is used to illustrate the impact of commodity divergences on factor prices. Figures 7.3a and 7.3b present input-output productivity curves for two industries that use the same input, unskilled labor. These industries are assumed to be the only employers of the factor. In industry 1, the marginal productivities of input use diminish sharply as increasing amounts of labor are applied in the production process. In industry 2, the marginal productivities are almost constant, giving the input-output productivity curve a nearly linear shape. In both cases, the firm-level productivity curves are drawn under the assumption that all other input levels are held fixed. Variations in the amounts of nonlabor inputs would generate productivity curves for labor that differ from those illustrated in the figure.

The demand of the firm for unskilled labor is determined by the profitability of input use. If labor exhibits diminishing marginal returns (as is assumed in the figure), the most profitable use of labor will result when the marginal value of production just equals the incremental cost of labor input: $(\Delta Q_1)(P_1) = (\Delta Q_L)(w)$. Rearranging these terms yields a relationship between the input-output productivity ratio and the input-output price ratio: $\Delta Q_1 / \Delta Q_L = w / P_1$. The slope of the input-output productivity curve $(\Delta Q_1 / \Delta Q_L)$ equals the input-output price ratio at the maximum profit point. The assumption of diminishing marginal productivities implies that $\Delta Q_1 / \Delta Q_L$ exceeds w / P_1 whenever input use is less than this profit-maximizing level.

The same assumption also implies that larger quantities of labor will be used by the firm as wages decline, yielding a firm (and industry) labor demand curve that is downward sloping, as shown in Figure 7.3c. The industry 2 demand curve (Figure 7.3d) is more elastic than the industry 1 curve because the marginal productivity of labor is assumed to change more slowly as increased amounts of labor are used. Aggregation of the industry demand curves D_1 and D_2 yields the market demand for labor, illustrated in Figure 7.3e. The equilibrium wage rate, w^*, results from the intersection of the market demand curve with the available supply of labor, Q^*. The allocation of labor between the two industries is indicated in the industry demand diagrams as Q_{L1}^* and Q_{L2}^*.

If the interactions between labor inputs and other inputs (whose use levels are also affected by changes in the output price) are ignored, the same marginal productivity curve can be used to determine input demand under varying output prices. If the output price increases from P_1' to P_1'', the firm-level labor demand increases from q^1 to q^2, and the marginal productivity of labor in industry 1 declines. Changes in the number of firms in each industry will also affect the factor demand curve. As P_1 increases, industry 1 expands and industry 2 contracts.

The net shift in aggregate labor market demand could thus be positive or negative, depending on the relative intensity of factor use in the two industries. In the example, industry 1 is more labor-intensive than industry 2, and total labor demand shifts outward, from D^* to D^{**}. The equilibrium wage rises from w^* to w^{**}. Relative to the initial position, labor use has increased in industry 1 (Q_{L1}^* to Q_{L1}^{**}) and declined in industry 2 (Q_{L2}^* to Q_{L2}^{**}). Marginal physical products in industry 2 have had to increase to justify the higher wage. If industry 2 were the labor-intensive industry, aggregate demand would shift backward, off-setting the wage-increasing impact of industry 1 expansion. Wages and marginal physical products in both industries would decline.

As the number of commodities is increased, the relationships between factor prices and commodity price divergences become more obscure. Divergences that increase commodity prices increase factor demands by firms; divergences that decrease commodity prices shift firm factor demands backward. Changes in the numbers of firms in each industry cause further back-and-forth movements in aggregate factor demand. If protection is not biased toward any factor (if, for example, capital-intensive industries do not receive more protection than labor-intensive industries), the output price divergences will not have a substantial impact on factor prices. The factor price effect of a divergence in one commodity market would be offset by the factor price effect of a divergence in another output of opposite factor intensity. But when output prices are biased by factor intensity, factor price effects can arise.

Indirect Effects: Macroeconomic Distortion

The effects on factor prices of macroeconomic distortions will be transmitted indirectly, through commodity market prices, or directly, through a change in the cost of capital resources. The distorting effects of macroeconomic policy can be represented by the government budget deficit identity

$$(G - T) = (S - I) + (F^{in} - F^{out}),$$

where $G - T$ is the deficit, $S - I$ is the difference between domestic savings and investment, and $F^{in} - F^{out}$ is net foreign-exchange inflows.

If a deficit is financed through borrowing from the domestic capital market, domestic interest rates become unduly high; producers in all industries reduce capital use to increase the marginal physical product of capital and justify higher capital costs. If a deficit is financed through foreign borrowing, an overvalued exchange rate results. In the long run, the misvaluation is transmitted uniformly to the prices of all factors. In terms of equation 4, the exchange-rate effect is equivalent to a uniform (and in this case, positive) tariff on all commodities. Marginal physical products remain unchanged, because all tradable industries increase demand simultaneously.

Indirect Effects: Input Substitution

Input substitution incentives arise if the elimination of divergences causes relative factor prices to change. The producer then has incentives to try to lower costs (and alter marginal physical products) by altering the combinations of inputs used in production. Figure 7.4 illustrates the range of producer responses to changes in factor prices. In Figure 7.4a, input choices are fixed. Only one combination of inputs can be used in production, L units of labor and K units of capital. Prices of labor and capital are initially w and r, respectively, yielding isocost line AB. If the price of capital increases, the new isocost line will be A'B'. But the tangency to the production isoquant will remain unchanged. The input combination that minimizes production costs is still L units of labor and K units of capital.

In Figure 7.4b, the options for input choice are increased. An infinite number of combinations can be used to produce one unit of output. Initial choices are represented in the diagram by L units of labor and K units of capital. A change in the factor price then results in a different least-cost input combination. By increasing the use of labor and reducing the use of capital, the producer is able to reduce the impact of factor price changes on production costs. Total production cost in Figure 7.4b is less than that in Figure 7.4a, because the increase in labor costs, $(L' - L)w$, is more than offset by the reduction in capital cost, $(K - K')r$. As shown in Figure 7.4b, isocost line A''B'' lies inside isocost line A'B'.

Analysis of input substitution becomes more complicated when more than two inputs are present; changes in the use of these other inputs will cause shifts in the capital-labor isoquant. Figure 7.5 illustrates this result. The input-output productivity curve for labor is initially represented by 0A. This curve is drawn under conditions of fixed levels of all other inputs. If alteration of the other input prices leads to changes in their use, the productivity curve for capital and labor can be affected. In the figure, the productivity curve for labor shifts upward, from 0A to 0B. Such a shift would occur, for example, in the event of a reduction in the price of fertilizer. The amount of labor required to produce one unit of output is initially L_A, but this magnitude declines to L_B as fertilizer use increases. If fertilizer use exerts a similar effect on capital productivity, the unit production isoquant for capital and labor will shift inward, toward the origin. Given constant values for w and r, the new combinations of labor and capital become L'_A and K'_A.

A different complication arises if the productivity curve for capital

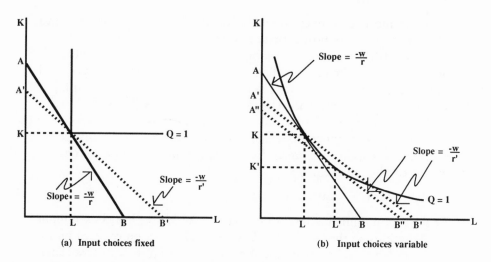

Figure 7.4. Input prices and input substitution

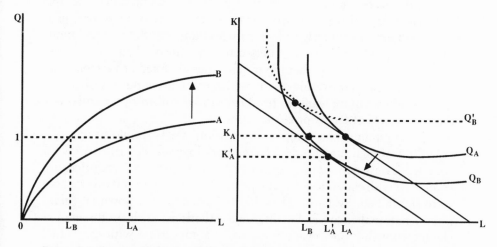

Figure 7.5. Input substitution with many inputs

shifts downward in response to increased fertilizer use, reflecting some strong complementarity between fertilizer and capital usage, so that the same quantity of output requires more, rather than less, capital. In this event, the new unit isoquant may be Q_B' instead of Q_B. Input substitution responses to a change in fertilizer use increase capital requirements and decrease labor requirements relative to the initial values of K_A and L_A. Although such complementarity effects exist, they do not appear so widespread as to dominate an economy's response to changing input prices. In most economies, input substitution relations are expected to reflect positive cross-productivity effects.

Input substitution creates changes in marginal physical productivities; when aggregated across all industries, the factor demand curve will shift, causing effects on factor prices. Without information about input substitution possibilities, the effects on factor prices will usually have to be ignored in empirical work. But encouragement for empiricists comes from the envelope theorem, which shows that the first-order changes in production costs are accounted for by changes in input prices. A change in cost can be represented as

$$(\Delta w)l + (\Delta r)k + w(\Delta l) + r(\Delta k) = \Delta C$$

At the cost-minimizing level of output, the producer has chosen input combinations so that $w(\Delta l) + r(\Delta k) = 0$. The effect on costs of an increase in labor use must be equal to the reduction in costs associated with a simultaneous reduction in capital use. If conditions were otherwise, producers could lower total costs by increasing labor use and decreasing capital use (or vice versa). At the margin, therefore, producers respond completely to factor price changes with "perfect" input substitution. The relationship between factor price changes and cost changes is the same, whether input quantities are fixed or variable. The intuitive appeal of the result is increased in the many-input case, because this case allows more ways for producers to substitute inputs and offset the cost effects of factor price increases.

In other circumstances, input substitution effects should be ignored, regardless of their magnitude. The methodology just described allows a complete assessment of the incentive effects of policy; private prices are compared with estimates of social prices that would exist if divergences were eliminated. Systems are evaluated in relation to potential (maximum) national income. But in some situations, such as foreign-exchange contributions, evaluations will be concerned with the actual contribution of the commodity system to national income. In the distorted economy, the opportunity costs of inputs to the system are determined by their social values in existing (distorted) production technologies. Thus second-best social factor prices would be calculated from world prices for commodities and the existing marginal physical productivities, allowing no role for input substitution effects.

Concluding Comments

Social valuation of domestic factors is the most difficult aspect of social cost accounting. The critical first step in estimating the social prices of factors is the development of a consistent framework in which to identify divergences. The exercise of quantification becomes a series of sequential adjustments to private market factor prices to recognize the effects of factor and commodity market divergences and the indirect effects of macroeconomic distortion and input substitution. As in all shadow pricing methods, complete knowledge of the response of commodity systems to price changes is necessary to derive exact estimates of social values.

Empirical estimates of social factor prices are thus approximations, and the analyst will be forced to make arbitrary judgments about what constitutes large and small changes. But so long as such judgments are made evident to others, better information or alternative ideas can be introduced to modify the results. The advantage of the approach developed here is its adaptability to different amounts of information. Because most of the potential errors in evaluation are introduced through their effects on domestic factor prices, only a small number of variables (perhaps only the wage rate and the rate of return to capital) need adjustment in the recalculation of social profitability.

Bibliographical Note to Chapter 7

Most research concerning the labor and capital markets has concentrated on supply behavior. The trade-theoretic approach ignores this source of variation at the aggregate level; all changes in factor supplies occur in the reallocation of a fixed total among alternative industries. This approach has allowed trade theory to concentrate on the effects of divergences in marginal value products among industries and divergences between marginal value products and factor prices. Three surveys of this literature are Stephen P. Magee, "Factor Market Distortions, Production and Trade: A Survey," *Oxford Economic Papers* 25 (March 1973): 1–43; W. M. Corden, *Trade Policy and Economic Welfare* (Oxford: Clarendon Press, 1974); and Corden, "The Normative Theory of International Trade," in *Handbook of International Economics,* ed. Ronald W. Jones and Peter Kenen (Amsterdam: North-Holland, 1986), 1, sec. 6, 63–130.

The relationship between commodity market divergences and factor shadow prices has been investigated in benefit-cost analysis; an increase in the shadow price of one factor relative to its market price implies that the price of some other factor has a shadow price below its factor price: Peter Diamond and J. A. Mirrlees, "Private Constant Returns and Public Shadow Prices," *Review of Economic Studies* 43 (February 1976): 41–48. But the empirical usefulness of this result depends on whether the activities that determine private market factor prices remain socially optimum. In the trade literature, similar results have been obtained in the course of generalizing the results of the Stolper-

Samuelson theorem. See Wilfred Ethier, "Some of the Theorems of International Trade with Many Goods and Factors," *Journal of International Economics* 4 (May 1974): 199–206; and Ethier, "Higher Dimensional Issues in Trade Theory," chap. 3 in Jones and Kenen, *Handbook of International Economics.*

Input substitution problems are examined intensively in the literature on the effective rate of protection and domestic resource cost. An early article is Wilfred Ethier, "Input Substitution and the Concept of the Effective Rate of Protection," *Journal of Political Economy* 80 (January/February 1972): 34–47. A survey is provided in sec. 3.1 of Ronald W. Jones and J. Peter Neary, "The Positive Theory of International Trade," in Jones and Kenen, *Handbook of International Economics.* The role of input substitution and differences between first-best and second-best shadow prices for factors is a principal focus of the papers by Ronald Findlay and Stanislaw Wellisz, "Project Evaluation, Shadow Prices and Trade Policy," *Journal of Political Economy* 84 (June 1976): 543–52; and T. N. Srinivasan and Jagdish N. Bhagwati, "Shadow Prices for Project Selection in the Presence of Distortions: Effective Rates of Protection and Domestic Resource Costs," *Journal of Political Economy* 86 (January 1978): 97–116. A somewhat heartening result for empiricists is that very large input substitution effects appear necessary to induce perverse results: Ronald W. Jones, "Effective Protection and Substitution," *Journal of International Economics* 1 (February 1971): 59–81; and Harry G. Johnson, "Factor Market Distortions and the Shape of the Transformation Curve," *Econometrica* 34 (July 1966): 686–98.

PART IV

EMPIRICAL ESTIMATION OF THE POLICY ANALYSIS MATRIX

CHAPTER 8

Constructing PAMs
for Commodity Systems

THIS CHAPTER presents a framework for empirical estimation of the PAM. Estimation of the elements in the first row of the matrix—private revenues, costs, and profits—is straightforward. The numbers in this row are based on direct observation of actual revenues and costs in existing commodity systems. But calculation of the second row—social revenues, costs, and profits—is more complex. The analyst is asked to evaluate the effects on production systems of policies and market failures, thus altering the observed values of revenues and costs. A further complication for empirical estimation arises because commodity systems are composed of a set of activities—production, processing, and marketing. These activities represent the unit of observation, and commodity systems are aggregations of the costs and returns to each activity. The following section elaborates on these aspects of the estimation exercise.

The chapter also considers issues in the organization and presentation of budget data. The first task for the development of the PAM is to select systems that are closely related to the policy issues of interest. In this identification process, decisions are made about farm production, movement of the commodity from the farm to the processor, processing, and transport to a wholesale market. Because the PAM uses both private and social prices for inputs and outputs, cost and returns information is disaggregated in two ways. First, quantity and unit price data are usually necessary for estimation of social costs and returns. Second, private costs are classified into four categories—labor, capital, land, and tradable inputs—so that the analyst can identify the impact of divergences on social costs of production.

The Estimation Strategy

The procedure entailed in the empirical construction of PAMs can be seen by rewriting the letter entries of PAM in terms of price and quantity variables. The PAM can be described as follows:

	Revenues	Tradable Inputs	Domestic Factors	Profit
Private	P^D	$\sum_i p_i^D q_i^D$	$\sum_j w_j^D l_j^D$	π^D
Social	P^S	$\sum_i p_i^S q_i^S$	$\sum_j w_j^S l_j^S$	π^S

where P = price of output, p_i = price of tradable input i, q_i = quantity of i per unit of output (Q), w_j = price of factor input j, l_j = quantity of j per unit of output, and π = profit. A superscript D is used to indicate that the value of the variable is observed under existing (private) price incentives; superscript S denotes the value that the parameter would assume under social price incentives. The above PAM describes costs and revenues as values per unit of output; the q_i and l_j represent input-output coefficients. But the matrix values can be equally well presented as values per hectare, values per firm, or in terms of any other unit of observation. The q_i and l_j need only to be multiplied by the relevant output measure.

To estimate PAMs, representative systems are first identified. Next, for each system, observable data for prices, output levels, and input use are collected, and the first line of the PAM is estimated. Third, the price and quantity observations are modified to reflect the social values appropriate to the second line of the matrix. The necessary social prices may be observed directly (world prices for tradable outputs and inputs) or they may be derived indirectly (for example, using information about divergences to estimate social factor prices from private factor prices). Finally, the observed quantities of inputs and outputs are altered to their "social" values, using econometric information about price response or engineering information about alternative technologies. If fixed input-output coefficients are assumed, the latter step is omitted.

The particular PAM that has been discussed so far represents revenues and costs for a commodity system—a chain of farming, processing, and marketing activities that characterize the production and delivery of a commodity to a wholesale market. But PAMs for commodity systems are not estimated directly. Instead they are composites of PAMs for each activity in the chain. For the purposes of data collection and organization, the PAM framework defines a commodity system to include four activities—farm production, delivery from farm to processor, processing, and delivery from processor to the wholesale mar-

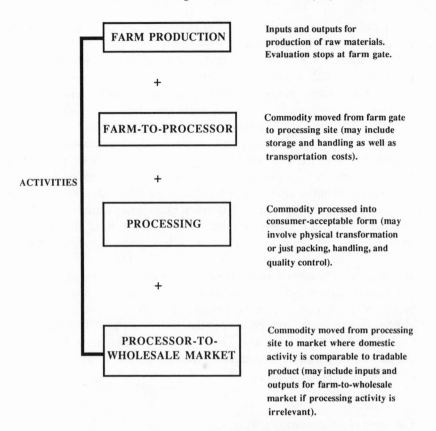

FARM PRODUCTION	Inputs and outputs for production of raw materials. Evaluation stops at farm gate.
+	
FARM-TO-PROCESSOR	Commodity moved from farm gate to processing site (may include storage and handling as well as transportation costs).
+	
PROCESSING	Commodity processed into consumer-acceptable form (may involve physical transformation or just packing, handling, and quality control).
+	
PROCESSOR-TO-WHOLESALE MARKET	Commodity moved from processing site to market where domestic activity is comparable to tradable product (may include inputs and outputs for farm-to-wholesale market if processing activity is irrelevant).

ACTIVITIES

Figure 8.1. The structure of the commodity system for PAM analysis

ket. Figure 8.1 illustrates the structure of the commodity system model. Each of the activities in Figure 8.1 is described by a PAM matrix, made up of price and quantity variables. The PAM for the commodity system is derived from aggregation of revenues and costs across the representative activities. In some cases, these elements cannot be directly added to one another but must be adjusted to avoid double-counting of revenues and costs. Summation of output revenues for each activity, for example, would involve multiple counting of the principal output.

Additional calculations are needed to adjust for differences in the commodity units or numeraries used by different activities. Cost and revenue data for the activity budgets are collected initially in whatever form is most convenient. Farm-level costs of wheat production are commonly estimated on the basis of land area (such as costs per hectare). Farm-to-processor costs, such as transportation services, are measured on a per metric ton (or some other weight or volume measure) basis. Budgets for processing and processor-to-market activities might use different numeraires as well.

Conversion to a common numeraire is achieved with conversion ratios. Figure 8.2 describes the adjustment process for a wheat flour production system. In the top half of the figure, the system costs and revenues are expressed in currency units per physical unit (Portuguese escudos per metric ton) of wheat flour, the final product of the commodity system. If farm activity costs and revenues are measured initially as currency units per land area (escudos per hectare), these entries need to be adjusted to final product equivalents. Two conversion ratios are necessary—the inverse of farm yield (hectares per metric ton of wheat) and the inverse of the processing outturn ratio (metric tons of wheat per metric ton of flour). When farm-level costs and revenues are multiplied by these two conversion ratios, the farm-level entries are converted to an escudos per metric ton of flour basis. For the farm-to-processor activity, only the inverse of the processing outturn ratio is needed as an adjustment factor. No adjustments are needed for the processing and processor-to-market activities, because these costs and revenues are already denominated in escudos per metric ton of flour. The choice of a numeraire is entirely arbitrary. The bottom half of Figure 8.2 illustrates the activity adjustment procedure for the calculation of system costs and revenues on a per hectare basis.

Defining the Commodity System

Every economic activity is unique in some way. In the farm sector, for example, commodity choices, land quality, and input use patterns are almost never identical for any two farms. Although it is possible to develop a different production model for every farm, such exercises are impractical because of limits on the resources available for research. They are also of no use to policy-makers, because the design of separate policies for each farm is impossible. Instead, decisions affect broad categories of farmers, defined in terms of geographic location, commodity produced, and technologies. Since no two farmers are affected in exactly the same way by a particular policy action, policy-makers usually base decisions on the average impact of the policy on some particular group of farmers.

Development of a list of potential representative systems and subsequent reduction of the list to a manageable number are thus the initial tasks in the construction of a PAM. The selection of commodity systems

Activity	Original units of measure for activity	Conversion ratios for activity and secondary product revenues	Adjusted units of measure for activity
Farm	Escudos/hectare	$\left(\dfrac{\text{Hectares}}{\text{Mt wheat}}\right) \times \left(\dfrac{\text{Mt wheat}}{\text{Mt flour}}\right)$	Escudos/mt flour
Farm-to-processor	Escudos/mt wheat	$\left(\dfrac{\text{Mt wheat}}{\text{Mt flour}}\right)$	Escudos/mt flour
Processing	Escudos/mt flour	None	Escudos/mt flour
Processor-to-market	Escudos/mt flour	None	Escudos/mt flour
Farm	Escudos/hectare	None	Escudos/hectare
Farm-to-processor	Escudos/mt wheat	$\left(\dfrac{\text{Mt wheat}}{\text{Hectare}}\right)$	Escudos/hectare
Processing	Escudos/mt flour	$\left(\dfrac{\text{Mt flour}}{\text{Mt wheat}}\right) \times \left(\dfrac{\text{Mt wheat}}{\text{Hectare}}\right)$	Escudos/hectare
Processor-to-market	Escudos/mt flour	$\left(\dfrac{\text{Mt flour}}{\text{Mt wheat}}\right) \times \left(\dfrac{\text{Mt wheat}}{\text{Hectare}}\right)$	Escudos/hectare

Figure 8.2. Conversion ratios and the calculation of system costs and returns

is perhaps the most arbitrary, yet crucial, element of PAM research. Characteristics that are similar across firms are chosen as the basis for the representative firm. Commodity produced, region of production, and production technology are the most common identification criteria. The choice of characteristic depends on the policy issue. If policy-makers are concerned about wheat price policy, a wheat commodity system will be needed. If the question is fertilizer pricing policy, one studies commodity systems that are the prominent users of fertilizer. Because the change in policy could alter the behavior of firms, poten-tially prominent commodity systems, as well as those that are currently in operation, must be anticipated.

While the analyst is searching for parameters that can serve to aggre-gate individual farms into larger representative categories, concern must be given also to aggregate farms that are very different in other respects. This simultaneous concern for similarity and diversity means that more than one representative system may be necessary for PAM analysis. For example, if wheat farms differ substantially in production technologies or soil qualities, a single commodity system will not be adequate to model the wheat sector. The analyst then needs to distinguish represen-tative wheat farms that are machinery-intensive, labor-intensive, or animal-intensive and that have good or poor soils.

Recognition of diversity in representative systems is constrained by the resources and time available for the research effort and by the differential impacts of policy on the systems. Since the preparation of budgets is a labor-intensive exercise, projects rarely have the resources to study more than fifteen or twenty representative systems. At the same time, the effects of a particular policy (or set of policies) might not differ across systems in a major way. If changes in the price of wheat affect profitability similarly across different types of irrigation technologies, for example, little insight will be gained from explicit models of ground-water pumping and gravity-fed irrigation systems. More important distinctions might involve irrigation versus rainfed technology. In this event, the representative systems could include groundwater pumping and nonirrigated-wheat production techniques.

In most instances, the costs and returns of each potential system are not known a priori. Consequently, judgments about representative sys-tems will be arbitrary, reflecting the ability of the analyst to anticipate the important and trivial differences in the results. As budgets are constructed and initial results evolve, differences between some repre-sentative systems might turn out to be small. If so, the list of representa-tive systems can be shortened. To ensure that systems encompass the full range of variation in costs, returns, and policy response, the initial sets of representative systems should be too large rather than too small. Box 8.1 illustrates the system identification and selection process for wheat in Portugal.

The representative commodity system includes more than just a farm-level production activity. Consideration of farm-level costs and returns would be sufficient to evaluate the efficiency and competitiveness of production for home consumption. But interest is more often directed to production for a domestic or foreign market that is geographically distinct from the farm. Selection of a representative market destination means that post-farm costs must be included in evaluation of the system. Furthermore, the more critical policy issues and incentive effects may be entailed in the post-farm activities. Critical constraints to increased farm production may be less related to farm technologies and farm policies, but instead a consequence of ineffective or excessively costly marketing activities. Policy analysis of post-farm activities then becomes more important than analysis of the farm level activity.

Activity selection is also dictated by the requirements of social evaluation. The domestically produced product must be comparable to a commodity available in international markets. For example, both wheat flour and wheat grain are available in world markets. As a result, analysts of representative wheat systems may choose to ignore flour processing altogether and emphasize system variations in wheat production, transportation, and storage activities. Alternatively, analytical interest might focus on wheat flour processing. The farm production activity could be omitted, and flour production systems could represent variations in processing technologies, transportation, and storage. In this example, wheat becomes a tradable input for the processing activity; its domestic market price reflects the miller's costs, and its social value is represented by the world market price plus the social costs of delivery to the mill.

Classification of Inputs and Outputs

The budget of output revenues and input costs provides the organizational framework for data collection and the construction of a PAM. A budget is constructed for each activity of the system. Data collection begins with compilation of an inventory of inputs and outputs for each activity. These items are categorized, quantified, and priced, first in private and then in social terms. The costs and returns of each activity

Box 8.1. Selecting Representative Wheat Production Systems in Portugal

Portuguese wheat production occurs in three agroclimatic zones: the Alentejo, a hot, dry, rainfed area with large, mechanized farms; the Ribatejo, a neighboring region with a somewhat cooler climate, better-quality soils, extensive irrigation, and a high degree of mechanization; and the Tras-os-Montes, a region with a cooler climate than that of the Alentejo, limited irrigation, and a range of mechanized and animal-intensive production technologies. Preliminary inspection of production systems in the various regions revealed a large number of potential representative systems. In the Alentejo, the most prominent differences in wheat production were those associated with the four soil qualities that are recognized in the soil type classification used in Portugal. In the Ribatejo, both rainfed and irrigated production technologies were present. Rainfed systems again differed by soil quality. Irrigated systems used either sprinkler irrigation or groundwater pumping into a furrow delivery system. In the Tras-os-Montes, production systems were primarily rainfed, differentiated by the use of animals versus tractors for land preparation and by soil quality. In total, twelve representative systems (four from each region) were considered as potential candidates for PAM analysis.

Only three systems actually were selected. How was this reduction achieved? In the Alentejo, the decision was made to model representative systems for high-quality (A–B) and low-quality (C–D) soils. In the Ribatejo, rainfed systems appeared very similar to the Alentejo systems; thus rainfed system models for the Ribatejo were judged redundant. Although irrigation technologies were different, preliminary analysis indicated that their effects on profitability were relatively minor. The decision was made to use the sprinkler irrigation technology as the representative irrigated-wheat system. The Tras-os-Montes technologies and profitabilities were quite different from those of the other regions, but the region's production was only a small percentage of total national output. For this reason, the Tras-os-Montes systems were left out of the final evaluation, allowing increased attention to data collection for the other representative systems.

Portugal joined the European Community in January 1986. The policy issue for wheat analysis involved assessment of the impacts of the Common Agricultural Policy (CAP) prices on Portugal's wheat sector. This assessment provided the rationale for ignoring the less productive region. But different policy issues could have generated a much different set of representative systems. If the concern had been low-income farmers, for example, the Ribatejo and much of the Alentejo would have been excluded, and principal attention would have focused on the Tras-so-Montes and parts of the poor-soil Alentejo. If the policy issue had involved subsidies for irrigation water, attention would have focused largely on the Ribatejo. There, distinctions between irrigation technologies would have been crucial in the designation of representative systems.

COSTS
    I. Fixed input
    II. Direct labor
    III. Intermediate input
    IV. Commodity-in-process

REVENUES
    V. Output

Figure 8.3. Input and output categories for activity budgets

are added together to generate the total costs and returns for the commodity system. Figure 8.3 contains a listing of the principal categories of inputs and outputs for each budget.

Fixed Inputs

Budgets are made up of costs and returns on an annual (or single crop) basis. However, fixed inputs have a useful life of many years, and only a portion of fixed input costs should be attributed to a particular year's production. A simple approach is to divide the cost of the fixed input by the useful life of the input. But that calculation ignores the need for capital expenditures to earn a rate of return on the investment. For example, if a wheat farmer did not buy a tractor, the money could have been invested in some other on- or off-farm activity. If this potential investment could earn a positive rate of return, the tractor investment must earn at least an equal return.

The annual equivalent value for a fixed input is known as the capital recovery cost—the annual payment that will repay the cost of a fixed input over the useful life of the input and will provide an economic rate of return on the investment. The derivation of capital recovery cost can be illustrated in a few steps. A is defined as the annual payment sufficient to repay the cost, Z, of the fixed input, at the end of its useful life of n years. If one puts amount A into an investment earning rate of return i, the total value of one's annual payments at the end of the fixed input's useful life will be

$$A(1 + (1 + i) + (1 + i)^2 + \ldots + (1 + i)^{n-1}) = Z \tag{1}$$

The term $A(1 + i)^{n-1}$ is the value of the initial deposit at the end of n years, the term $A(1 + i)^{n-2}$ is the value of the second deposit at the end of n years, and so on until the end: the term $A(1)$, which represents the value of the nth year payment. This formula calculates the amount of capital necessary to repay the cost of the fixed input.

Because the fixed input is required also to earn a positive rate of return, the necessary value of the output produced by the fixed input is not just Z but $Z(1 + i)^n$. Therefore, the annual cost-equivalent calculation is expressed by

$$A(1 + (1 + i) + (1 + i)^2 + \ldots + (1 + i)^{n-1}) = Z(1 + i)^n \qquad (2)$$

This expression can be altered by rearrangement of the terms to

$$A = (Z)(1 + i)^n/(1 + (1 + i) + (1 + i)^2 + \ldots + (1 + i)^{n-1}) \qquad (3)$$

Equation 3 can be written as

$$A = Z\left[\frac{(1 + i)^n i}{(1 + i)^n - 1}\right] \qquad (4)$$

The bracketed term on the right-hand side of equation 4 is the capital recovery factor. By applying this factor to the purchase price of the fixed input, the analyst can calculate the annual equivalent value for any fixed input.

Annual equivalent values also depend on initial capital cost (Z) and useful life (n). Replacement cost is used as the estimate of initial capital cost to maintain consistency with the long-run perspective of the PAM. Existing firms utilize many different vintages of capital equipment; as a result, fixed costs may vary substantially among firms. But capital stock must be replaced eventually, and current costs of fixed inputs become important to the continued operation of the firm. Useful lives of fixed inputs vary among firms as well, depending on intensity of use as well as owner maintenance practices. Equipment dealers and construction firms can be good sources of information about useful life. Rough rules of thumb can be used when no better information is available: buildings, 30 to 40 years; machinery, 10 to 15 years; and small machines and tools, 5 years. Box 8.2 provides some examples of the calculation of annual equivalent values for wheat production inputs in Portugal.

Direct Labor

The second category of inputs, direct labor, covers all labor directly employed in the activity. Both hired and family labor resources are included. If the analyst wants to make a distinction between family and hired labor, these inputs can be entered as separate lines within the labor

Box 8.2. Calculating Annual Equivalent Values of Fixed Inputs

The calculations needed to determine the annual equivalent value of a fixed input are slightly more complicated than is indicated in the text. First, the fixed input may have some salvage value when its useful life is ended. Because salvage value is realized at the end of the useful life, estimates of salvage value are discounted to the present before the net total cost of the fixed input is determined. These calculations are illustrated here for a tractor, tractor accessories, and a thresher in wheat production in Portugal. The rate of return used is 2 percent.

Input	Initial cost (thousands of escudos) (a)	Useful life (years)	Salvage value (thousands of escudos) (b)	Present value of salvage value (thousands of escudos) (c)	Net initial cost (thousands of escudos) (a − c)
Tractor	1,366.70	10	268.00	219.85	1,146.85
Plow	122.50	15	24.78	18.41	104.09
Disk	217.30	15	43.56	32.36	184.94
Planter	397.30	15	82.00	60.93	336.37
Thresher	3,482.10	10	374.10	306.89	3,175.21

A second complication for the calculation of annual equivalent values arises when the fixed input serves a larger number of units of the activity than are covered in the budget. A farm-level budget, for example, may be expressed in terms of costs per hectare, whereas the fixed input serves much more than one hectare during each year of its useful life. Only a portion of the annual equivalent costs of the fixed input should be allocated to the per hectare budget. These allocations are determined here for the five fixed inputs used in wheat production.

Input	Hours per hectare (d)	Hours per year (e)	Per hectare share of annual use (d/e)
Tractor	10.1	1,000	0.0101
Plow	3.0	250	0.0120
Disk	2.0	250	0.0080
Planter	1.0	125	0.0080
Thresher	2.0	400	0.0050

The annual capital cost per hectare is determined by the product of the net initial cost, the capital recovery factor, and the per hectare share of annual use, as is shown here.

Input	Net initial cost (f)	Capital recovery factor (g)	Share of annual use (h = d/e)	Annual capital cost (f g h)
Tractor	1,146.85	0.111327	0.0101	1.290
Plow	104.09	0.077825	0.0120	0.097
Disk	184.94	0.077825	0.0080	0.115
Planter	336.37	0.077825	0.0080	0.209
Thresher	3,175.21	0.111327	0.0050	1.767

category. Similar distinctions may be made between male and female laborers and laborers of different ages and skill levels. Again, separate line entries within the direct labor category provide a way to maintain an advantageous organization of the data.

The category does not include all the labor used by the system, because some labor will be indirectly employed as a consequence of the use of intermediate inputs by the activity. If a farm activity uses imported fertilizer, for example, the labor used to handle and transport the fertilizer to the farm gate is an indirect employment effect of the farm activity. Keeping separate the direct labor inputs facilitates the analysis of employment effects of the system; this topic is often of interest in policy debates about particular commodity systems.

Intermediate Inputs

The third category of inputs, intermediate (variable) inputs, are characterized by a useful life of less than one year, viewed from the perspective of the representative firm. Examples of items included in this category are seeds, fertilizer, pesticides, fuels, and lubricants. But also relevant are the rentals of capital equipment services, such as custom plowing, custom harvesting, pesticide application, and transportation services. Ultimately, many of these costs will be categorized as capital costs, but the valuation and analysis of these costs will often be different from those used in the evaluation of fixed inputs owned by the activity.

Commodities in Process

The final category of input is commodities in process. This category covers the commodity of interest for the PAM and is included only as an accounting convenience. Because profitability is calculated for each of the activities, the principal output of the commodity system appears several times in the budgets of the system. In a wheat flour system, for example, wheat is the principal output of the farm activity. Wheat is the commodity in process for the farm-to-processor activity and the processing activity. Wheat flour is the commodity in process for the processor-to-market activity. Inclusion of the wheat cost within each activity is necessary for calculation of activity profits. In the transport of wheat from farm to processor, for example, the purchase price for wheat at the farm gate and the sales price for wheat at the processor's mill are observed (or imputed). Similar calculations follow for evaluation of the processing activity (buying wheat and selling wheat flour) and the processor-to-wholesale-market activity (buying wheat flour from the mill and selling wheat flour at the wholesale market point).

In evaluation of the system costs and revenues, however, only the incremental costs of production from the postfarm activities are counted. Wheat flour production costs are the farm-level costs of wheat production; the transportation, handling, and storage costs of the farm-to-processor activity; the processing costs net of wheat for the processing activity; and the transport, handling, and storage costs of the processor-to-wholesale-market activity. If the analyst were to include the wheat costs in the aggregation of postfarm activity costs, the system costs would include two units of wheat and one unit of wheat flour, whereas system revenues would show one unit of wheat flour. Such calculations are clearly erroneous. When a separate category in each postfarm activity is kept for commodities in process, the relevant costs are easily aggregated to the system-level evaluation.

The input categorization process described here is not intended to be rigid. Alternative categorizations of inputs may be better suited to particular evaluation problems. Whether the input is included in the direct labor category or in the intermediate input category makes no difference to the final result. The critical aspect of the evaluation process and the construction of PAM is that all inputs should be included somewhere in the budgets. Without a comprehensive account of inputs, production costs are underestimated, biasing the results in favor of more positive private and social profitabilities.

Outputs

Revenues, the final category of the activity budget, cover all outputs of the activity. The commodity of interest is designated the primary output and is listed first in the category list. All other outputs are called secondary outputs and are entered on subsequent lines of the list. These designations are entirely arbitrary; secondary products can be as important as or more important than the primary product as sources of revenue for the activity.

The categorization reflects the particular focus of the research project. Whether or not the outputs are marketed makes no difference to the budget calculation. The distinction for a valuable output is that it has some productive value to the activity. This value can be gained through sale or use elsewhere in the activity. Meat production systems, for example, often generate manure that is used to fertilize crops. The budget for meat production must impute a value to the manure, so that

total revenues properly reflect the value of meat (and manure) production to the farm.

Evaluation of Inputs and Outputs

After inputs and outputs for each activity have been identified, they
need to be evaluated. The chosen time frame in which to evaluate the
costs and benefits of the activities is termed the base year for PAM
analysis. The base year may be the current year or any past year.
Research objectives and practical considerations determine the choice
of base year. If current policy issues are the focus of research, the base
year will be as near the present as possible. But current-year data,
especially price data, will often be incomplete, so the data must come
from one or two years in the past. Because policy-makers may be wary
of dated results, relevance to current issues requires the use of a base
year as close as possible to the present year. Alternatively, if historical
issues are of particular interest, the base year or years could go far into
the past.

Both quantity and unit price information for the estimation of costs
and returns are desirable to facilitate social valuation. The most common procedure used in the estimation of social input cost or social
output value is to apply social prices for inputs or outputs to the relevant
quantity measure. For some inputs, quantity and unit price data cannot
be isolated. In this circumstance, social values are approximated by
proportional adjustment of the private value of a particular input or
output. Sometimes information about divergences can be used to generate estimates of social values. For example, if the farm wheat budget has
only a total cost for pesticide input, without any indication of the
quantities used, information on the percentage distortion of import
prices can allow the researcher to impute a social value. If tariffs on
imports of pesticides are 50 percent, the private value of pesticides is 50
percent higher than the social value. Division of the total private value
by 1.5 gives an estimate of the social value. Such procedures entail the
assumption that quantities are unaffected by the price change.

When proportional adjustments to private values are impossible,
equality between private and social values is often presumed. If the
input or output accounts for a small proportion of total input costs or
output revenue of the activity, little harm is done to the results. Even if
the assumption of equal private and social values is incorrect, incorporation of the "true" social value will have an insignificant effect on

the magnitudes of total social costs and social revenues. But if the item in question is a large component of costs or revenues, the assumption of convenience can prove a grave error in practice. At this point, further analysis must be postponed until a more comprehensive set of data can be assembled.

Explicit recognition of the time frame of analysis provides another justification for the collection of separate price and quantity estimates for the major inputs and outputs of the system. From the policy-maker's perspective, the long-run profitability of the system is often most germane to the policy formation process. Because many policies are not changed with great frequency, the policy-system interaction over a long time period must be understood. In the portrayal of the longer-run interactions of policy and profitability, expected prices replace prices observed at a particular time as the correct measures for calculation of input costs and output revenues.

Disaggregating Input Costs into Domestic Factor and Tradable-Input Components

After all private and social input costs have been standardized to an annual basis, they are allocated to their domestic factor and tradable-input components. This disaggregation is necessary to permit identification of tradable-input and domestic factor divergences. Figure 8.4 illustrates the complete organizational format for the activity budgets. Both total private and total social costs are decomposed into their domestic factor and tradable-input components. In principle, many classes of domestic factors could be recognized. But for most purposes, four categories of domestic factors—unskilled labor, skilled labor, land, and capital—are sufficient. Because the commodity-in-process category is used only as an accounting device in the construction of the commodity system model, only the first three categories of input costs are disaggregated.

The decomposition exercise could be applied to every input listed in the fixed input, direct labor, and intermediate input categories. For example, the cost of fixed inputs reflects some marketing margin in addition to the basic cost of the machine. This margin incorporates the payments to factors and tradable inputs needed to operate the retail shop. Payments to hired labor could implicitly include payments for transportation to the activity site. Like the marketing margin, transportation costs reflect payments to a range of domestic factors and tradable inputs.

COSTS

	Quantity	Private costs and returns						Social costs and returns					
		Unskilled labor	Skilled labor	Land	Capital	Tradable inputs	Total	Unskilled labor	Skilled labor	Land	Capital	Tradable inputs	Total
I. Fixed input													
II. Direct labor													
III. Intermediate input													
IV. Commodity in process							$Z1$						$Z2$
Total		$C1$	$C2$	$C3$	$C4$	B		$G1$	$G2$	$G3$	$G4$	F	

REVENUES

	Private	Social
V. Output	A	E
VI. Profit	D	H

$$(D = A - [B + C1 + C2 + C3 + C4] - Z1)$$

$$(H = E - [F + G1 + G2 + G3 + G4] - Z2)$$

Figure 8.4. The structure of activity budgets for the PAM

Decomposing all input costs into their exact domestic factor and tradable-input components is a formidable task that can absorb substantial resources. Moreover, adjustment often will have only a trivial effect on the results. The noncapital cost components of fixed inputs and the nonlabor cost components of direct labor inputs are usually a very small proportion of total costs. Unless information about decomposition is readily available, fixed input costs are usually classified entirely in the capital cost category and direct labor inputs are classified entirely in the categories of unskilled labor and skilled labor.

In practice, the exercise is generally limited to the intermediate inputs. Again, on the basis of available information and resources for the research effort, many intermediate inputs can be classified into a single domestic factor or tradable category. Seeds, fertilizer, and pesticides are examples of intermediate inputs whose costs reflect marketing margins in addition to the pure tradable cost. But if these margins are judged to be relatively small, costs of intermediate inputs can be allocated exclusively to the tradable category without causing major errors in the results.

With other intermediate inputs, including electricity, transportation, and most services, no particular cost category appears to dominate total costs. Such inputs are denoted as nontradable inputs, because they are not available on international markets. The decomposition of these inputs implies the construction of an activity budget for production of the intermediate inputs that is as complicated as the one in Figure 8.4. Electricity production could be analyzed as an activity, for example, with a budget that identifies the fixed inputs, direct labor, and intermediate inputs necessary to produce electricity. In the process of decomposing the input costs for electricity production, more nontradable inputs—for example, machinery service and repairs—would be encountered. A budget could be constructed for the service and repairs in order to determine the proper allocation of these costs among the domestic factor and tradable categories.

Such calculations can take the analyst away from the original purpose—turning all policy analyses into studies of the nontradable industries in the economy. If these inputs are relatively unimportant elements of the commodity system costs, substantial research resources would be expended with little effect on the results. A rule of thumb is that unless the nontradable input represents more than 5 percent of total production costs of the system, separate budgeting exercises should be avoided.

More rapid approximations of the decomposition of nontradable inputs can be obtained in two ways. The most common technique utilizes an input-output matrix of the national accounts. These aggregate portraits of the economy allow calculation of the shares of labor and capital in each sector of the economy. Land costs typically are ignored, because they are a small component of nontradable-goods production costs. When the nontradable input of interest is associated with a particular sector, the capital and labor cost shares can be approximated. The remainder is allocated to tradables. This exercise provides a decomposition of the private costs of the nontradable input. Social costs of the nontradable input are then estimated by adjustment of the labor, capital, and tradable components to reflect the impacts of divergences. The sum of the social values of the domestic factor and tradable components gives the total social value of the nontradable input.

A second alternative for the treatment of nontradable inputs relies more on the analyst's judgment. When input-output matrices of the economy are unavailable, the distribution of costs among domestic factor and tradable categories must be estimated. In the absence of any information, an operational rule for distributive shares is the assumption that nontradable inputs contain one-third labor, one-third capital, and one-third tradables. Each private cost component is then adjusted to its social value, and the social values of the labor, capital, and tradable components are summed to generate an estimate of total social cost. Such estimation exercises are not much different from pure guesswork. If these arbitrary calculations are commonplace in the system evaluation, more data collection is essential before the system analysis should proceed. Box 8.3 illustrates the decomposition of nontradable inputs for a wheat system in Portugal.

Concluding Comments

Because the PAM is an accounting framework composed of identities, users may choose among empirical techniques to estimate its elements. The budget-based approach described here has shortcomings. As this and subsequent chapters make clear, budgets place heavy reliance on judgment and informed guesswork by analysts. Statistical measures to indicate representativeness are often unavailable, output supply and input substitution responses to social prices are often incorporated in an approximate manner, disaggregations of input costs comprise many approximations, and fixed input-output coefficients are sometimes a necessary assumption. But budget-based approaches have great advantages as well. The data are relatively easy to collect and do not depend on the long time series that so often confounds econometric estimation.

Box 8.3. Decomposition of Nontradable Inputs for Wheat Production Systems in Portugal

Decomposition of intermediate inputs for good-soil wheat production in southern Portugal, 1981 (in escudos per hectare)

Intermediate inputs	Quantity (kilogram)	Decomposition of private costs					Total private costs	Decomposition of social costs					Total social costs	Net policy effects
		Tradable inputs	Labor	Land	Capital	Total		Tradable inputs	Labor	Land	Capital	Total		
			Domestic factors						Domestic factors					
Seeds	170	3,910.00	—	—	—	—	3,910.00	3,910.00	—	—	—	—	3,910.00	0.00
Disinfectants	0.3	60.00	—	—	—	—	60.00	60.00	—	—	—	—	60.00	0.00
Fertilizer	250	2,192.50	—	—	—	—	2,192.50	4,982.65	—	—	—	—	4,982.56	−2,840.06
Fuel and lubricants		1,952.15	448.03	—	800.06	1,248.09	3,200.24	1,600.12	373.36	—	1,201.04	1,547.40	3,174.52	25.72
Machinery repairs		1,718.17	394.34	—	704.17	1,098.51	2,816.68	1,408.34	328.62	—	1,057.09	1,383.71	2,794.05	22.63
Building repairs		3.05	.070	—	1.25	1.95	5.00	2.50	0.58	—	1.88	2.46	4.96	0.04
Custom spraying		1,235.25	283.50	—	506.25	789.75	2,205.00	1,012.50	236.25	—	760.00	996.25	2,008.75	16.25
Totals		11,021.10	1,126.57	—	2,011.73	3,148.30	14,159.40	12,976.12	938.81	—	3,020.01	3,958.82		−2,775.42

Note: (—) signifies not applicable.

This table contains 1981 data on intermediate inputs into a good-soil wheat production system located in southern Portugal. The initial data are the costs per hectare for each intermediate input, shown as total private costs. The intermediate inputs are then classified as either tradables or nontradables. Three items—seeds, disinfectants, and fertilizer—are classified as tradables. The net policy transfer for fertilizer is a large subsidy (shown as − 2,840.06) that reduces the private costs of farmers.

The other four intermediate inputs are classified as nontradable. The task is to decompose each of the four nontradable intermediates into tradable inputs, labor, and capital. Three principal distortions—a fuel tax, a labor tax, and a capital subsidy—affect the private prices of these nontradables. Only one distorting commodity policy, a 22 percent tax on fuel, affects the price and use of inputs into the nontradables. The private price of labor (market wage costs per hectare) is judged to be higher than it would be in the absence of policy because of legislation requiring vacation bonuses and social security contributions that do not have to be paid by other employers. Accordingly, all private labor costs exceed social labor charges by 20 percent. Government policy on capital is designed to create a subsidy that is enjoyed by wheat farmers and their input suppliers; the annual market interest rate is taken as 2 percent (in real terms, after correcting for inflation), compared with an estimated shadow real interest rate of 8 percent.

The net effect of these three kinds of policies on the four nontradables, shown in the right-hand column of the table, is a slight tax. When both tradable and nontradable intermediates are considered, the substantial subsidy on fertilizer (− 2,840.06) swamps the small net tax on nontradables (64.64) and creates a net policy transfer (− 2,775.42) that subsidizes 16 percent of the total social costs of intermediate inputs into wheat farming.

The use of a disaggregated framework of costs simplifies the introduction of more accurate information in a piecemeal fashion. Finally, the data can be selected to correspond to a number of research issues: the effects of policies on particular commodities, regions, or types of producers; the attractiveness of alternative technologies; or the effects of variations in access to input and output markets.

Bibliographical Note to Chapter 8

Much of the work on budget-based estimation techniques has been provided in the context of linear programming. A classic reference on this subject is Robert Dorfman, P. A. Samuelson, and R. Solow, *Linear Programming and Economic Analysis* (New York: McGraw-Hill, 1958). Further discussion of the use of linear approximations of economic behavior is provided in John Duloy and Peter Hazell, "Substitution and Nonlinearities in Planning Models," in C. R. Blitzer, P. B. Clark, and L. Taylor, eds., *Economy-Wide Models and Development Planning* (New York: Oxford University Press, 1975), pp. 307–25. Linear models have been used to analyze all levels of economic activity. The work by Blitzer, Clark, and Taylor contains discussions of national aggregate models. An example of agricultural sector analysis is Roger D. Norton and Leopoldo Solis M., *The Book of CHAC: Programming Studies for Mexican Agriculture* (Baltimore: Johns Hopkins University Press, 1982). But the most frequent application of programming models has been at the level of the firm. Examples of this work include Carl Gotsch et al., "Linear Programming and Agricultural Policy: Micro Studies of the Pakistan Punjab," *Food Research Institute Studies* 14 (1975), pp. 3–105.

The preparation of budgets requires few resources other than a hand calculator; spreadsheet programs of micro-computers can also be useful for developing budget formats and automating many of the calculations. A reference work that provides capital recovery factors and other measures that depend on interest rates is J. Price Gittinger, ed., *Compounding and Discounting Tables for Project Evaluation* (Baltimore: Johns Hopkins University Press, 1973).

CHAPTER 9

Farm-level Budgets and Analysis

FARM-LEVEL ISSUES receive most attention from agricultural policy-makers. Ministries of agriculture usually maintain cost-of-produc-
tion estimates for principal commodities. But ministries focus most
often on private profitability, whereas the PAM analyst is concerned
also with estimation of transfers induced by policy or market failures.
These differing analytical objectives place new demands on the existing
data. For example, the ministry requires only the total costs of intermediate inputs, whereas the PAM analyst is concerned with the price
and quantity used of each input in order to measure the effects of price
distortions or to assess the potential impacts of input substitution.
Consequently, in most instances, substantial data gathering efforts will
be necessary to permit the construction of PAMs for representative farm
activities.

This chapter considers the more common strategies and problems
encountered in using farm-level budget data to prepare a PAM. Policy
issues dictate choice of crops, level of aggregation, and selection of
indicators to measure representativeness. The chapter next discusses the
complementary uses of budget data, national or regional production
data, and experiment station (or demonstration farm) data. Because of
time and cost constraints on the research project, the principal role for
farm-level field work is the verification and modification of secondary
data and the collection of appropriate private market prices. The most
difficult pricing exercises usually involve primary inputs (especially
labor and capital) and nonmarketed goods. Finally, the chapter considers some of the complications that arise in trying to portray farm
decisions in the framework of a single commodity system. Because

many farmers produce multiple crops, the disaggregation of fixed inputs in particular commodity systems may require some arbitrary assumptions. Intercropping, agronomic constraints on crop rotations, and perennial crops present further complications for the calculation of private and social profits.

Selection of Representative Crop Activities

Choices of farm activities are determined by the research problem and the scope of agricultural issues identified by the government. For example, if policy-makers are interested in the tax/subsidy impact of government policies on the agricultural sector, one or two representative budgets for each crop should be sufficient. If the research instead focuses on a single crop or technology, a more detailed specification of commodity production is needed and a larger number of representative firms should be used. Sectoral income distribution objectives require commodity systems that highlight the small farm–large farm dichotomy; concerns for regional growth require recognition of region-specific commodity systems.

Within each group of commodity systems, analysts may still desire to characterize production heterogeneity in some detail. Regional classifications are perhaps the most common indicator of heterogeneity, because differences in agroclimatic zones—characterized by soil fertility, topography, and access to water—typically influence the choice of technology and the level of input use. Differences in farm size are a second source of heterogeneity. Small farms often use variable inputs, such as fertilizer and labor, with different intensities than large farms. Large-farm systems are often capital-intensive, and fixed costs account for a more substantial share of total costs. Differential access to resources creates heterogeneity in the relative usage of machinery and labor inputs.

The development of a list of representative systems can draw on various information sources. Aggregate production estimates, usually prepared by the ministry of agriculture, can be obtained for particular crops and regions. Sometimes these estimates are decomposed by farm size or technological characteristics (number of animals or amount of machinery, for example); in this circumstance, aggregate data can be used to specify the technological alternatives as well. But the identification of specific technologies usually requires first-hand observation and the assistance of farm management personnel from development projects, universities or ministries of agriculture, and members of local extension services. Initial lists of potentially representative systems for PAM analysis will usually be longer than the research projects can manage, and some reduction in the list will be necessary. Short field trips with expert observers are useful at this stage to give the analyst a better idea of the distinctions among commodity systems. Box 9.1 illustrates the system selection process for agriculture in northwest Mexico.

Procedures for Budget Preparation

Once representative firms have been identified, the estimation of budgets can proceed. Because PAM results are adversely affected by the omission of cost or revenue items, budgets should reflect a complete set of input and output activities. Preparation of the cropping calendar—a time line that identifies the various tasks in crop production, such as land clearing and preparation, planting, fertilization, pest control, and harvesting—reduces the likelihood of data omissions. This information is often readily available from extension agents or secondary sources. Otherwise, primary data on cropping calendars are relatively easy to collect. A single visit to each type of farm is usually sufficient to gain an adequate picture of cultivation practices.

The next step in budget preparation involves the specification of inputs and outputs associated with each task of the cropping calendar. Outputs are placed in a single category; inputs are classified into fixed, direct labor, and intermediate inputs. Fixed inputs include only the capital equipment owned by the operator of the farm activity. Direct labor inputs of both family and hired workers are maintained in a separate category for reasons of convenience. The employment effects of the activity and changes in assumptions about labor requirements are commonly of interest to the analyst, and separate categorization makes such calculations readily accessible.

Because each input will be evaluated in social as well as private prices, inputs have to be identified with a high degree of specificity. Labor is often divided into four categories—unskilled adult male, unskilled adult female, unskilled child, and skilled—because these types of labor usually have private market wages and social opportunity costs that differ from one another. Machinery such as tractors, plows, harrows, planes, wagons, and seed planters needs to be identified individually as well.

The cropping calendar approach might overlook some infrastructural

Box 9.1. Identification of Agricultural Commodity Systems in Northwest
Mexico

A research project in northwest Mexico had two principal objectives. The
first was to assess the international competitiveness of Northwest irrigated
agriculture, a sector dominated by capital-intensive techniques and reputed to
be dependent on numerous input subsidies. The second objective was to evaluate the impact of government policy on the profitability of the *ejido*, a small-
farm system established by the government. The evaluation was made through
comparison of the policy incentives for private farms with the incentives provided to the *ejidos*.

A vast number of farm systems were available for analysis. The two largest
agricultural states of the northwest—Sonora and Sinaloa—contain twelve irrigation districts. More than forty different crops are grown. Almost all of these
crops are found on both *ejiditario* and private farms. In total, hundreds of
commodity systems are present.

To reduce the set of commodity systems to a manageable number, aggregate
production data were examined. The set of crops was reduced to fourteen. The
prominent staples were corn, wheat, beans, sorghum, and rice. Oilseeds included soybeans, safflower, and sesame. The principal vegetables were tomatoes (large, salad, and cherry), green peppers, and potatoes. Cotton, chickpeas,
and alfalfa were also prominent crops. Site visits sugggested relatively little
difference in technologies among irrigation districts. This factor allowed reduction of the number of irrigation districts to four, two in Sonora and two in
Sinaloa. To allow consideration of the subsidies to different irrigation techniques, one district was chosen because it utilized groundwater pumping,
whereas the remaining districts utilized dams and gravity-fed water supplies. In
total, 120 commodity systems were identified.

The number of systems is large, but substantial economies of scale were
present in the data collection and analysis. For a given crop, principal technology differences among irrigation districts were usually limited to type of irrigation. In many cases, the *ejido* and private-farm techniques were identical;
differences were primarily in yield, because of differential land and management quality. Postfarm technologies were limited to transportation and storage,
and these technologies were similar for many of the commodities. Many elements of activity budgets, therefore, were transferable across representative
systems. But even with substantial amounts of secondary data on costs and
returns, data collection required about six months. In many policy analysis
situations, time constraints and the lack of secondary data will limit researchers
to far fewer systems. Between ten and twenty systems is a more common range,
and usually several months of effort will be required for construction of the
activity budgets.

inputs, such as barns, silos, and primary irrigation works, necessary to farming but not directly involved in the field processes of production. In most studies, only a portion of the costs of infrastructure will be attributed to the activity budget. For example, if the budget concerns costs and returns for one hectare of wheat, the analyst might choose a proportion factor, such as the inverse of farm size, as the share of the cost of the infrastructural input to allocate to the activity budget. The choice of a proportion factor is arbitrary. Infrastructural inputs are indivisible fixed costs for the farm, whereas the budget calculations require that farm costs be allocated among various cropping activities. No correct allocation exists for such inputs. Because such inputs are indivisible, one activity may contribute more or less than others to the costs of infrastructure.

The amount of particular inputs or outputs associated with the farm activity must be consistent with the choice of farm-level numeraire, usually a unit of area (acres or hectares, for example) or, for animal production systems, a specified herd size. For most fixed and direct labor inputs, quantity measures will be readily available. But in some circumstances, they will be hard to obtain, and the analyst will have only total cost for an input or total revenue for an output. These cases arise most frequently with intermediate inputs and secondary outputs. For example, the analyst could have estimates of machinery service costs or of pesticide and pesticide application costs. But quantity measures—hours of machinery time and kilograms or liters of pesticide— might not be available.

The absence of quantity measures is not a problem for PAM construction, so long as the social values can be calculated from the estimates of private market costs or returns. If social value can be estimated in terms of a percentage premium or discount relative to private value, a separate quantity estimate is unnecessary. When the particular input or output is a small conponent of total costs or revenues, arbitrary estimates of percentage differences between private and social values can be used to complete the budgets with relatively little effect on the results. But if the items are important elements of total costs or revenues, the arbitrariness of such estimates becomes a critical element in the reporting of results of the analyses. Policy-makers cannot evaluate results without some notion of the reliability of the estimates of costs and returns.

For inputs and outputs that are identified in quantity terms, unit prices represent the final ingredient necessary for the formulation of the budget. All prices need to be standardized to a common time period.

When prices are not from the existing time period, the analyst can impute them from available data by applying an inflation adjustment. Prices must also be standardized for location. To calculate the farm-level profitability of the activity, farm-gate prices or price equivalents are the relevant values. For intermediate inputs, prices therefore include marketing costs incurred in delivering the input to the farm. For example, the cost of fertilizer is not the ex-factory price but the ex-factory price plus the costs of marketing and delivering the fertilizer to the farm gate. Outputs should be valued similarly—not with the price in some consumer center but with a price or price equivalent that represents the ex-farm-gate value.

For direct labor, intermediate inputs, and outputs, the quantity and price information is sufficient to calculate private cost. As described in the previous chapter, the valuation of fixed and capital equipment inputs requires additional information on useful life and salvage value. Capital recovery factors are applied to determine annual equivalent costs of the fixed inputs. The costs are then multiplied by the share of total annual use (for example, hours per hectare divided by hours per year) to derive the fixed input cost for the activity budget. Figure 9.1 provides a summary of the quantity and price data most commonly needed for farm activity budgets.

Collection of Input and Output Data

Agricultural production is characterized by large numbers of firms at dispersed locations. In most cases, farms lack formal records of input use, particularly with regard to individual crops. Output records are somewhat more common, but usually this information is not expressed in the yield measures needed for economic analysis. As a result, primary farm surveys are expensive and time-consuming and place heavy demands on skilled manpower for monitoring and evaluating the survey data. In PAM-related work, the constraints of time and financial support for research usually mean that primary farm surveys are not possible. Instead, the analyst relies on secondary data in the preparation of representative farm budgets. Fieldwork remains critical to the construction of the PAM, but efforts focus on the verification of secondary data, the collection of information about current prices, and the introduction of modifications of input-output relations to account for technological change.

In most circumstances, prior data on farm budgets are available. The

Figure 9.1. Inputs and outputs in farm activity budgets

Category	Quantity measures per numeraire	Price measures
Fixed inputs: buildings, fences, land development	Useful life	Purchase price
Investments, irrigation infrastructure and equipment, machinery, machinery accessories, tools, work animals	Share of annual use	Salvage value, rate of return
Direct labor: unskilled male, female, child; skilled labor, by task	Days or hours	Wage per day or per hour
Intermediate inputs: seeds, fertilizers, insecticides, custom machinery services, repair and servicing of equipment	Weight or volume; most service charges not quantifiable	Farm-gate price per unit
Outputs: main products	Weight or volume	Farm-gate price per unit

Figure 9.1. Inputs and outputs in farm activity budgets

ministry of agriculture, producer organizations, and university researchers in agricultural economics often produce farm budgets, and their surveys can provide estimates of input and output quantities. Agricultural investment project proposals require economic feasibility analyses; estimates of farm-level costs and returns are usually included in this work. Extension service personnel might also have useful information about the input and output quantity requirements for particular commodity systems. If this information is not recorded in reports, it can usually be collected through visits to the agent. Finally, studies of comparable technologies in neighboring countries sometimes provide useful farm budgets.

Whatever the source of budget information, fieldwork usually begins with interviews of the employees who originally prepared the budgets. Such interviews are useful to disaggregate information about costs and returns beyond the level provided in published documents, to assess the extent of heterogeneity of production practices and the need for multiple budgets, and to gain initial impressions about the price and quantity effects of particular policy distortions or market failures faced by producers. Field informants might also arrange interviews with farmers and other informed observers of the local agricultural economy, such as providers of input or marketing services. Interviews provide supplementary information about prices and the efficiency of various input and output markets. They can also cover the particular policy issues that motivate the research. Because the selection of expert informants is not random, care must be exercised in using responses to characterize the various representative systems. But this approach has the advantage of confining fieldwork to several weeks rather than many months.

Quantities

Output data (crop yields or animal productivity) may be obtainable from ministry statistical branches responsible for national production estimates. If these data are available as a regional time series, the analyst can obtain useful estimates of normal yields. Because yields reflect economically influenced levels of input use and agronomically influenced varietal performance, care must be taken in designating particular yields as normal. Similar considerations require caution in the use of experiment station data. Under experimental conditions, the profitability of production is usually irrelevant and cultural practices that maximize yield rather than economic efficiency are the norm. Experiment station yields thus commonly overestimate on-farm yield levels. Associating these yields with estimates of on-farm costs causes profitability estimates to be excessive as well.

Experimental data can be useful, however, in estimation of the relative advantage of a new variety or cultural practice. Experimental plots are often used to compare new practices to the traditional practice in a control plot. The application of relative premia to actual yield data from farm surveys gives an estimate of the expected on-farm yields from the improved practice. This calculation presumes that the new technology will exert similar effects on both control plots and actual farms. If the control plot and actual farm yields are very different, the presumption may be erroneous. In this circumstance, only replication of on-farm practice can indicate the likely yield benefits from new technologies.

Different sources of secondary information on input and output quantities almost always show some differences in estimates. Comparisons of secondary data can be assisted by use of cropping calendars, which list the alternative estimates of input and output quantities for each farming task. Comparisons with other information sources, such as crop yield surveys, and the results of field interviews can then be used to make judgments about the quality of information from each source.

If the differences among estimates reflect variations in survey quality (caused, for example, by small sample size or careless survey design), the poorer-quality estimate can be disregarded. But estimates may also differ because the commodity systems are not the same. If the differences in estimates arise because of variations in local economic conditions or technologies, the description of the representative commodity system must be made more explicit. For example, if comparisons of fertilizer use estimates reveal one high estimate and one low estimate, these differences might be explained by differences in the sizes of farms sampled in the two surveys. Explicit decisions then need to be made about farm size in the description of the commodity system.

From careful comparisons of secondary information sources, a synthetic representative budget is constructed. This budget may use different sources of information for quantity estimates of each particular input and output. For example, one study could provide a particularly convincing estimation of average yields; another might be judged superior for its measurement of direct labor inputs; yet another could be the source of the most accurate measurement of water use and irrigation practices. The chief danger of synthetic constructions is that estimates from different technologies may be unwittingly blended to create a budget that is not representative at all. A second problem for synthetic budgets occurs when input measures are not consistent with output measures. The "best" estimate of fertilizer use might come from a study that showed relatively high yield estimates. Combining the fertilizer estimates with national or regional average yield data might result in a "representative" budget that overestimates fertilizer input relative to output. Profitability in the PAM calculations would be underestimated. Again, field visits and consultation with expert observers become necessary to verify consistency among the input and output measures of the representative budgets.

If secondary data for input and output quantities are absent, PAM analysis usually is not possible. Research resources must be devoted to the collection of such information rather than the compilation of budgets. But even in these circumstances, the analyst might be able to construct a representative commodity system with relatively little primary survey work. In this approach, secondary input data from available commodity systems are used as benchmarks for the estimation of input requirements of other commodity systems. Interviewers ask farmers or other experts for information about labor utilization and intermediate input use relative to the requirements of alternative commodities that have well-understood input-output relationships, assuming a plot of equal size for each commodity. This information is relatively easy to collect. With it, one can apply appropriate discounts or premia to the estimates from the alternative crops, providing a budget for a new commodity.

Prices

An equally important aspect of farm-level fieldwork involves collection of prices for inputs and outputs. Secondary data sources often provide some price data. Statistical offices frequently collect annual market prices of principal agricultural outputs, and secondary sources of budget data contain price data for inputs and outputs. A problem with direct use of these data arises when the base year for PAM analysis differs from the year used for the data from secondary sources. In addition, these prices might not represent expected market prices but instead might be the outcomes of peculiar demand and supply conditions.

For some inputs and outputs, market prices will not exist because the product is produced and consumed exclusively on the farm. These situations are particularly common in subsistence-oriented areas, where inputs such as manure and forages might never be traded on markets. In this situation, one needs a market-equivalent value for the product—the price at which the product would sell if a market existed. In many circumstances, this valuation can be based on comparison with a substitute commodity sold through markets. For example, animal feeds can substitute for forages, and the number of feed units contained in forage can be calculated and evaluated at the market price of a feed unit for animal feeds. Because substitution is rarely perfect—for example, animal feeds might not contain the roughage provided by forages—the search for market-equivalent values will often be an exercise in approximation. When substitute inputs are not available, market-equivalent values have to be estimated on the basis of the labor, capital, and intermediate inputs required to produce the input. The total costs of these inputs are assumed to reflect an implicit market price for the product.

Perhaps the most common nonmarketed input is family labor. Instead of receiving a wage payment, a family laborer shares in the net income of the farm. Each family member receives an implicit wage equal to the value of individual consumption and savings divided by the time devoted to the farm activity. Makers of budgets usually avoid such calculations by applying market wages to all labor inputs. If family labor does not earn the market wage (private profit is negative), at least some family workers could do better financially by leaving their own farms and seeking employment as hired laborers. The analyst then needs to develop a rationale for acceptance of a relatively low rate of remuneration, such as limited alternative employment opportunities or a desire for food security and a consequent unwillingness to rely on markets for basic foodstuffs.

This treatment is not entirely satisfactory. As discussed in earlier chapters, implicit wages would ideally reflect private marginal products, and divergences of the sort just described would become part of the explanation for differences between private and social costs of labor. But because family labor wages cannot be observed, market wages become a necessary substitute.

The determination of private market wages can be a complicated task. Nonmonetary incentives, such as meals or drinks, are often provided by employers. Because these items are a cost to the activity, the market-equivalent values of nonmonetary incentives are included in the calculation of labor wage rates. Market wages also reflect differences among family members in skill level, sex, and age, making it unlikely that a single wage rate will apply to all the labor inputs described in the budget. During slack seasons of the crop production cycles, wages might fall to a subsistence level or, in the event of a total absence of labor demand, temporarily to zero.

Complex Commodity Systems

In budget calculations, the relationship between output and the activity numeraire is expressed as the output of a single crop per unit area, such as kilograms of wheat per hectare. In most instances, the crop can be planted every year, using the same or equivalent production technology. The data collection exercise focuses only on the inputs and the yield for a single commodity.

Multiple Commodities

Sometimes the cropping patters will not be so simple. In intercropping, for example, two or more crops are grown simultaneously on a particular parcel because of some mutually beneficial relationship, such as reduced likelihood of pest problems (vegetables and staple food crops) or more efficient use of space (grass for animal feed intercropped with vineyards). A second type of complexity arises when agronomic considerations require crop rotation on a particular parcel of land. Some crops, such as cotton, place substantial demands on soil and usually are rotated with alternative crops to maintain soil fertility.

Models of multiple commodity situations can choose between two alternatives. In the sustainable unit area approach, a representative hectare (or acre, feddan, or other area-based numeraire) includes all agricultural practices required by the representative crop system. For example, if cotton cultivation is limited by agronomic constraints to two of every three years, a representative hectare includes two-thirds of a hectare of cotton and one-third of a hectare of an alternative crop. Intercrop systems will be based on shares of area occupied by the various crops.

The second alternative is the pure-stand-equivalent approach. The commodity system is modeled as if only one commodity were grown on the land. Input and output measures for the pure-stand-equivalent budget are estimated by division of observed data by the decimal share of the commodity in the mixed-crop system. For example, if corn yields in a mixed-crop system are 1 metric ton per hectare and the corn occupies only two-thirds of the hectare (the remainder being devoted to other commodities), the pure-stand-equivalent yield is $1/0.67 = 1.5$.

The choice between the two alternatives is dictated by the availability of data and research resources and by the representativeness of the farm system. The sustainable unit area method is more data-intensive, because inputs, outputs, and prices for each crop must be estimated and then summed in proportion to their relative importance in the cropping pattern. But this more complicated approach might be necessary, particularly if the analysis requires estimates of policy effects per unit area or per farm. Only the sustainable unit area approach can reflect the interplay of agronomic constraints and PAM results. Calculations of private profit, social profit, and divergences for a crop rotation, for example, are required, because farmers subject to rotational constraints engage in diversified cropping rather than complete specialization.

If agronomic constraints are not binding, economic analysis of the commodity system can ignore the complications of multicrop systems and utilize the computationally easier pure-stand-equivalent approach. Rotational constraints are irrelevant if land is in excess supply, because the producer can choose to leave it fallow at necessary intervals rather than altering the crop mix. If crop interaction effects have little impact on yields or levels of input use, the phenomenon of multiple crops is again uninteresting from the PAM perspective. In this situation, the choice of the farmer to grow multiple commodities in an intercrop fashion or in pure stands makes no difference to input-output relation-ships, and the simpler pure-stand approach suffices for the construction of PAM budgets. Other commodities may be ignored; only the share of the targeted commodity per unit area is needed to estimate the budget for the pure-stand equivalent.

Permanent Crops

Permanent crops, such as tree crops or vineyards, present another group of problems for budget estimates. A sustainable unit area model could be built, so that the representative area included different stages of the crop life cycle. Each year of a ten-year crop cycle, for example, is represented in one-tenth of a unit area in the representative crop budget. The inputs and outputs from each portion are then added to give total output and input requirements for the sustainable unit area. The result-ing production pattern is sustainable over an infinite time. The problem with the sustainable unit area calculations for permanent crops is the omission of the time-related costs of production. In the example of a ten-year crop cycle, the sustainable level of profits indicated by the budget will not actually be achieved until eleven years after the project's inception. But the sustainable unit area method calculates profits as if they were available every year, from the inception of the activity.

Only a project cycle evaluation approach can provide an exact eval-uation of benefits and costs that vary over time. In this procedure, a budget is prepared to represent each year of the crop cycle, with annual equivalent costs for all inputs and outputs. Each year's budget is thus prepared in a manner identical to that used for annual crops. Revenues, costs, and net profits from each year are then discounted to a present value and summed to indicate the expected present value of the use of the land over the cycle. Division of total present values by the number of years in the cycle determines annual average costs and revenues. In most projects, benefits are relatively small in early years and relatively large in later years. Discounting the time path of net benefits reduces future values more than early period values, thus yielding lower totals for the project cycle method than for the sustainable area method. Box 9.2 compares the two approaches to the evaulation of permanent crops.

Many permanent crops have a long production cycle; for example, most tree crops and vineyards have useful lives of 20 to 30 years. In this case, an alternative presentation of results may prove convenient. The budget represents the observed costs and returns of the activity in a year of full production (commonly attained six or seven years after initial

164 Empirical Estimation of the Policy Analysis Matrix

Box 9.2. Calculation of the PAM for Permanent Crops

The nut crop is assumed to have a three-year life cycle. The per hectare quantities and prices for inputs and outputs in each year are described in the following tables.

Year 1: Inputs and outputs

			Useful life (years)	Share of annual use	Annual cost per hectare
Fixed inputs	*Initial cost*	*Salvage value*			
Tractor	$10,000	$0	15	.04	$ 38.54
Plow	1,000	0	20	.04	3.21
Weeding tools	250	0	5	1.0	57.74
	Quantity	*Unit price*			
Direct Labor					
Skilled	2 days	$50.00			100.00
Unskilled	50 days	20.00			1,000.00
Intermediate Inputs					
Fertilizer (urea)	100 kg	0.25			25.00
Seedlings	1,000	0.50			50.00
Outputs					
None					

Year 2: Inputs and outputs

			Useful life (years)	Share of annual use	Annual cost per hectare
Fixed inputs	*Initial cost*	*Salvage value*			
Weeding tools	$250	$0	5	1.0	$ 57.74
Sacks	100	0	2	1.0	53.78
	Quantity	*Unit price*			
Direct Labor					
Unskilled	100 days	$20.00			2,000.00
Intermediate Inputs					
Fertilizer (urea)	200 kg	0.25			50.00
Outputs					
Nuts	500 kg	1.50			750.00

Year 3: Inputs and outputs

			Useful life (years)	Share of annual use	Annual cost per hectare
Fixed inputs	*Initial cost*	*Salvage value*			
Weeding tools	$250	$0	5	1.0	$ 57.74
Sacks	800	0	2	1.0	430.24
	Quantity	*Unit price*			
Direct Labor					
Unskilled	150 days	$20.00			3,000.00
Intermediate Inputs None					
Outputs					
Nuts	4,000 kg	1.50			6,000.00

The time path of total costs and revenues is shown here.

Year	Undiscounted costs	Undiscounted revenues	Costs (discounted at 5 percent)	Revenues (discounted at 5 percent)
1			$1,213.80	$ 0
2	$1274.49	$ 0	1,960.56	680.27
3	2161.52	750	3,013.05	5,183.03
Total	3487.98	6,000	6,187.41	5,863.30
Annual average costs per hectare			2,062.47	1,954.43

The last line indicates the numbers for use in the PAM. The actual calculations needed for the PAM would be more complex than what is shown here, because separate entries would be needed for labor, capital, and tradable inputs and because all inputs would be evaluated in social prices as well as in private prices.

The sustainable unit area method would ignore the time path of costs and benefits and would estimate per hectare values as follows.

Area	Costs	Revenues
⅓ hectare (year 1 type)	$ 424.83	$ 0
⅓ hectare (year 2 type)	720.51	250.00
⅓ hectare (year 3 type)	1,162.66	2,000.00
Total (1 hectare)	2,308.00	2,250.00

This method overestimates profit, because crop production has more costs relative to revenues in the early periods.

planting). Net profitability from each of the previous years is calculated and then compounded to give a net present value in the year of full production. These present values are summed and treated as an investment cost, where the useful life of the investment is the remaining term of the production cycle. For example, if a coffee production cycle were 30 years, reaching full production in the eighth year, the activity budget could represent the eighth year. The value of net profits in each of the first seven years is compounded to give present value in the eighth year. Profits from the first seven years are then summed and amortized over the remaining 22 years using the capital recovery cost method. This procedure generates an annual equivalent cost of investment that can be added to the representative budget as a part of fixed costs.

Technological Change and Partial Budgeting

The construction of representative budgets for specific crops is time-consuming, even when secondary data are available to provide most of the quantity and price information. Once the budget is constructed, the marginal costs of further use and modifications of budget data are relatively small. The numerous variations of a representative budget can be generated easily by alteration of a subset of input and output data. This exercise is termed partial budgeting.

Partial budgeting is most often used in the PAM methodology as a means of assessing the effects of new technologies on farm profitability. A new seed-fertilizer package for rice, for example, would be modeled by alteration of a traditional technology budget for changes in seeds, fertilizer, and yield. If the new technology increases yields, the budget might have to be modified further to recognize additional labor requirements for tending and harvesting the crop. Although such procedures seem mechanical, they are often useful portraits of the actual process of technological change. Farmers rarely jump from one set of practices to a new technology that uses entirely different inputs and practices. Instead, they modify current practices to incorporate a particular innovation.

The input and output data required for partial budgeting ideally are drawn from observations of the actual practice of farmers. Even if producers do not know specific quantities of inputs used and outputs derived, estimates can be obtained with comparative questionnaire techniques, in which producers provide information about the performance of the new technology relative to the old one. If information about actual practice is not available, the analyst is forced to rely on experiment station results or to modify them to reflect expected farm practice.

Comparisons of old and new budgets give the analyst information about the economic incentives for technological change. Consideration of both profitability and changes in the structure of costs is necessary for this assessment. Even if the new technology proves more profitable than the old one, potential constraints to adoption could appear. Cash-flow problems sometimes arise when new technologies entail a greater use of purchased inputs. The lack of marketing services can also limit adaptation. If marketing boards must handle the increment in production induced by technological change, physical and financial facilities might have to be expanded. By aggregating the representative budgets to a regional or national level (for example, multiplying per hectare budget data by the number of hectares on which the technique is used), the analyst can generate the total impact of technological changes.

Partial budgeting techniques can also be used in formulation of the agenda for future research and development. New technologies, such as high-yielding seed-fertilizer packages, improved means of pest control, substitutions among machinery and labor, and better water control and management, are often a direct consequence of the pattern of investment in research. Working with technical experts, agricultural economists can use partial budgeting techniques to simulate the impact of hypothetical technological changes on profitability. If potential changes do not create positive private and social profits and improvements over traditional techniques, alternative investment paths or changes in policies need investigation.

Whole Farm Analysis

Crop-specific budgets indicate the profit incentive to produce a particular commodity. But they give no insights into the contribution of particular commodity systems to farm income, the income characteristics of the farm operation, or the presence of input constraints on the expansion of particular systems. If representative budgets are available for all the principal farm crops, a representative farm can be modeled as a weighted sum of individual crop budgets, where the weights are units of the numeraire (such as number of hectares or number of animals per farm). Measures of total input and output requirements are determined by addition of the weighted sums across the individual crops. Such exercises add size of farm and crop mix to the list of factors that must be specified in the identification of a representative farm. Census data, advice from extension agents, and casual observation are usually sufficient to characterize typical types of farms.

These calculations give additional insight into data quality. The addition of input requirements across crops provides estimates of total demand for each input. Comparisons of aggregate labor requirements with labor supply (family plus hired) give an indication of the reliability of estimated labor requirements, for example. Whole farm models also avoid the arbitrary assumptions that may be needed to generate models of single commodity systems. Fixed inputs do not need to be allocated among commodities, and their costs are determined by total use (on-farm use plus rental of the inputs to others).

The main attraction of whole farm analysis is insight into farm income. Three income-related issues are amenable to PAM analysis. One is the estimate of total income and consideration of the consumption opportunities afforded by particular farm sizes. Some division of labor and capital costs is made to differentiate between owned inputs and rented inputs (hired labor and machinery rental). The sum of own-input costs and profits, less land rental payments (if any), gives a measure of the income received by the household from farming activities. Estimates of off-farm income are then added to give total farm family income. Second, if whole farm models are developed for various sizes of farms, the analyst can draw inferences about the bias (if any) that a policy demonstrates toward particular types of farms. Finally, whole farm models offer insights into the dynamics of technological change by estimating the potential change in total income from a particular innovation and the capacity of individuals to self-finance new investments in land, machinery, or other improved inputs. Income less consumption yields savings, and whole farm budgets can be combined with information about consumption requirements to generate measures of potential financial contribution of the farm to new investments. Such calculations yield insights into the importance of credit markets and imperfections that distort the access to credit.

Concluding Comments

To prepare farm budgets, policy analysts need complete familiarity with production systems. Arbitrary decisions are made at all stages of the research effort, from the selection of representative systems to decisions about normal levels of quantity and price. These judgments usually must be made without the comfort of statistical tests for representativeness. In these circumstances, time spent in fieldwork becomes essential for the development of budgets. The great attraction of the budget-based approach is the complementarity between secondary data and fieldwork, allowing field time to be measured in weeks instead of years. Even senior analysts are afforded the opportunity to view and understand production systems directly rather than having to interpret them through the eyes of enumerators or the vagaries of a massive data base. Such close association greatly enhances the potential for relevant policy analysis.

Bibliographical Note to Chapter 9

Much of the literature about data collection procedures and strategies is unpublished or has received only limited circulation. Perhaps the most widely available source dealing with farm budget compilation is Maxwell L. Brown, *Farm Budgets: From Farm Income Analysis to Agricultural Project Analysis* (Baltimore: Johns Hopkins University Press, 1979). The organization of farm budget information is discussed in C. Peter Timmer, Walter P. Falcon, and Scott R. Pearson, *Food Policy Analysis* (Baltimore: Johns Hopkins University Press, 1983), chap. 3. Two other documents, reflecting substantial experience with field surveys and interview techniques, have been published by the International Maize and Wheat Improvement Center (CIMMYT), *Planning Techniques Appropriate to Farmers: Concepts and Procedures* (Mexico City: CIMMYT, 1980), and Richard K. Perrin et al., "From Agronomic Data to Farmer Recommendations: An Economics Training Manual" (Mexico City: CIMMYT, 1976).

One criterion by which to identify representative systems is regional variation. The existence of intraregional differences with respect to farm size remains a controversial empirical issue. Most explanations for farm size differences rely on the presence of an imperfect labor market that causes systematic variation in the opportunity costs of this input. The issue is discussed in R. Albert Berry and W. R. Cline, *Agrarian Structure and Productivity in Developing Countries* (Baltimore: Johns Hopkins University Press, 1979). Additional explanations of differential productivity, related to the availability of credit and the supervision of hired labor, are developed in Gershon Feder, "The Relation between Farm Size and Farm Productivity," *Journal of Development Economics*, 18 (August 1985): 297–313.

Other applications of farm modeling, such as partial budgeting, are discussed in Brown, *Farm Budgets*. An application of partial budgeting to the analysis of technological change is provided in Charles P. Humphreys and Scott R. Pearson, "Choice of Technique in Sahelian Rice Production," *Food Research Institute Studies* 17 (1979–1980): 235–77.

PAMs are most commonly used for evaluating single commodity systems, and the farm is assumed to be made up of separable crop production functions. But decisions about one crop may affect decisions about another crop; this interdependence is termed jointness, and its implications are explored in Richard

Just and David Zilberman, "Multicrop Production Functions," *American Journal of Agricultural Economics* 65 (November 1983): 780–90; and C. Shumway, R. Pope, and E. Nash, "Allocatable Fixed Inputs and Jointness in Agricultural Production: Implications for Economic Modeling," *American Journal of Agricultural Economics* 66 (February 1984): 72–78. Whole farm models of farms are described in J. Price Gittinger, *Economic Analysis of Agricultural Projects,* 2d ed. (Baltimore: Johns Hopkins University Press, 1982), chap. 4. An application of whole farm analysis that utilizes the PAM approach is provided by Roger Fox and Timothy J. Finan, "Patterns of Technical Change in the Northwest," in Scott R. Pearson et al., *Portuguese Agriculture in Transition* (Ithaca: Cornell University Press, 1987), pp. 187–201.

CHAPTER 10

Postfarm Budgets and Analysis

POSTFARM ACTIVITIES of agricultural systems are economic functions—transportation and handling, storage, processing, and sales—that link farmers with consumers in domestic or international markets. Because the competitiveness of production agriculture can be measured only at the point of consumption, postfarm activities are an essential influence on private and social profitability. Sometimes postfarm costs are more important than farm production costs in the determination of the final consumer price and system efficiency.

Budgets for postfarm activities are critical also to understanding the price formation process. Accurate measurement of marketing costs and returns provides insights into the competitiveness of various stages of marketing. Analysis of postfarm budgets can suggest ways that governments might narrow margins, thus raising farm-gate prices relative to consumer market prices. Evaluation of postfarm activities is important also in understanding the reasons for use of particular policy instruments. Agricultural price policy objectives usually are pursued indirectly through the determination of price at some point away from the farm, such as at a consumer market or a storage facility. These prices are then transmitted back to the farm through the marketing system, with each stage of the marketing process commanding some portion of the policy price.

This chapter discusses procedures for the construction of postfarm budgets. Postfarm data gathering follows a process similar to that used in farm budget preparation. Descriptive analyses of marketing chains precede the selection of representative firms. Because of the large number of activities, some elements of the marketing system will receive less attention than others. The dearth of secondary data for budget prepara-

171

tion implies a heavier reliance on primary surveys than with farm budget preparation.

Selection of Representative Postfarm Activities

The identification and selection of representative marketing activities proceeds in a manner analogous to that used in construction of farm budgets. Initial efforts identify the principal marketing chains that carry output from the farm to the consumer. Visits to principal production and processing areas and conversations with commodity experts are usually sufficient to identify the alternative methods of marketing and processing. These chains vary in complexity from simple storage on the farm for home consumption to complex systems of farm-gate collection, bulking in local markets, transportation to main consumption centers, processing, and distribution to consumer markets. Because analytical interest centers on the competitiveness of the domestically produced commodity relative to potential competing imports or exports, the marketing chain is usually terminated at the final wholesale point. Both imports and domestic production pass through the wholesale-retail linkages, and costs and returns at this stage of marketing thus have no influence on the relative competitiveness of the two products. If the commodity is an export, marketing chains are terminated at the fob point, where the product is ready to leave the local port for foreign delivery.

The actual number of marketing chains will generally exceed the number that can be analyzed effectively. Market shares for each marketing chain can be used to reduce the set of marketing activities. However, some marketing chains may be unimportant under observed market conditions but potentially important once the effects of divergences are removed. In addition, the policy issues of interest could mandate attention to particular marketing chains precisely because they are unimportant; the issue in this case is to identify why those marketing chains are little used. Box 10.1 illustrates the system identification and selection process for rice marketing in Ghana.

A key consideration in choosing postfarm activities for evaluation in the PAM is the relative importance of marketing costs in the agricultural system. If marketing costs command only a minor share of the final market price (at either the domestic market or the export point), a single representative marketing chain is often sufficient for construction of the PAM. Data collection and analysis can focus instead on variations in

Box 10.1. Alternative Marketing Chains for Delivering Rice to an Urban Market, Atebubu District, Ghana

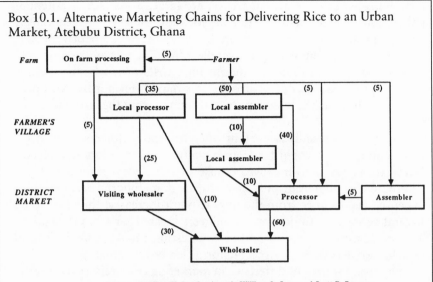

Source: Based on V. Roy Southworth, William O. Jones and Scott R. Pearson, "Food Crop Marketing in Atebubu District, Ghana", <u>Food Research Institute Studies</u> 17(1979): p. 171.

In this example, rice moves to the urban market in seven ways. All rice must be processed before consumption. Three processing technologies—hand pounding, processing in small-scale mills in the local villages, and processing in a large-scale mill in the major town of the district—were observed. Bulking agents, or assemblers, were also found to operate at various levels—buying directly from the farmer in both the local village and district markets and buying from one another and assembling progressively larger volumes of rice before sale to the large processor in the district market.

If the research project were focused on farm production issues, only two marketing chains would be selected for analysis: local processing at the village and distribution through a visiting wholesaler; and bulking through a local assembler, large-scale processing in the district market, and shipment to the wholesale market. In Atebubu District, Ghana, these two systems accounted for about 85 percent of marketed output. The hand-pounding technique was judged unimportant for marketed output (although it was more significant for home consumption), and the other marketing chains differed little from the selected systems in terms of physical input use.

Different policy concerns would mandate the inclusion of additional marketing chains. Interest in employment effects would require consideration of the relatively labor-intensive chains, such as the hand-pounding and multiple assembly chains. An interest in comparing the competitiveness of rice production against imports would require analysis of marketing chains for three areas of consumption—on-farm consumption, local village consumption, and consumption in major cities.

farm budgets. But if marketing costs are large, a rich set of potential policy issues emerges. In particular, location, marketing technology, and end product become important determinants of competitiveness. A lack of competitiveness might not reflect high farm-level costs of production but instead might be due to high costs of marketing. In this event, representative commodity systems may differ not by farm production technique but by location, marketing technology, or end product.

Location is especially important when transportation costs are large. Production areas located far from potential export points will receive farm-gate prices that are low because of substantial transport costs. Alternatively, high transport costs mean that imports become more expensive as distance from the import point increases, thus providing natural protection to domestic producers located far from that point. Transportation cost issues become interesting cases for analysis, and representative systems are specific for different locations.

The second source of difference in marketing costs relates to alternative marketing technologies. Differing technologies can be present in every phase of the marketing process, but they are usually of greatest significance in storage and processing. In developing country agriculture, large losses and high interest costs mean that storage costs can be substantial. Although all storage technologies have interest costs, differences in the magnitude of losses sometimes lead to substantial differences in carrying costs. Even more significant variations are found for processing. For example, techniques for processing paddy rice vary from hand pounding with a mortar and pestle to processing a few hundred kilograms per hour with small mills to dehusking, cleaning, and bagging several tons per hour with large mills.

End products provide a final characteristic in the selection of multiple marketing systems. Some farm products can be transformed into a variety of commodities for consumption. For example, corn can be eaten directly as human food, consumed as flour, fed as grain to animals, incorporated into animal feed, processed into corn oil or corn starch, or processed into sweeteners or alcohol. If the farm product is traded on world markets (as corn is), the choice of end product is an issue more for industrial development than for agricultural development. Corn prices are independent of the domestic processing industries. But when the farm product is not traded internationally (raw milk, for example), choice of output (cheese, butter, skim milk powder, or whole milk) can be an important influence on potential farm-gate prices and profitability. Competitiveness depends on the domestic efficiency of all activities in the commodity system.

Procedures for Budget Preparation

A budget can be formulated wherever a market transaction occurs. This approach provides a maximum disaggregation of the marketing system and allows detailed comparisons between costs and margins for each stage of the marketing process. But if the number of budgets is large and at least some of the margins are small, a simplified analysis can consider fewer budgets. The PAM format described here uses only three levels of postfarm budgets: a farm-to-processing activity that includes transportation, bulking, and storage of the farm product; a processing activity that includes processing costs and possibly storage of both the processed and farm products; and a processing-to-market activity that includes transportation and delivery costs to the wholesale point.

Postfarm budget data utilize numeraires different from those of the farm production activity. Often, the numeraires differ among marketing activities. A typical rice system, for example, would have farm-to-processor costs denominated per metric ton of paddy rice, whereas processing and processor-to-market costs would be measured per metric ton of milled rice. Conversion ratios—number of kilograms of milled rice per kilogram of paddy rice (that is, milling outturn ratios)—are necessary to convert each marketing budget to a common numeraire before the total costs of the rice marketing system can be calculated. The conversion ratios can affect substantially the importance of particular activities in total system costs, and become prime candidates for sensitivity analyses.

Farm-to-Processor and Processor-to-Market Budgets

The postfarm activities other than processing are involved primarily in transportation and storage. The technologies of these activities are usually easy to describe, and relatively few inputs are involved. Labor needs include unskilled manual labor (handlers) and skilled labor (drivers, managers of warehouses, and merchants). Fixed input requirements are limited to warehouses, trucks, and machinery for loading and unloading. Intermediate input costs include working capital (to represent the opportunity cost of storage), fuel, maintenance and repairs on transportation equipment, and sacks or other handling materials. A final cost element, losses for commodities while in storage, can vary widely, depending on the commodity characteristics and storage technology. Unless the storage agent keeps records of warehouse throughput, estimates of losses can only be approximated.

The costs of transportation activities depend importantly on the analytical treatment of capacity utilization and transport distance. Because fixed costs are spread over annual throughput, some estimates of annual activity (number of trips per year, average number of tons stored per year) are necessary for the calculation of fixed costs per unit of product. Distance is determined by the designation of location in the representative budget. But knowledge of the rate structure of transport costs is useful when the PAM results are generalized to locations other than those specified in the representative budget. Transport costs generally vary less than proportionally with distance, reflecting economies of size and one-time costs of loading.

Time is also a critical parameter, particularly for the estimation of storage costs. The principal components of storage costs—losses and the opportunity cost of financial capital invested in the stored commodity—are directly related to the duration of storage. If the commodity is stored for a one-month period, the warehouse operator will seek to recover the purchase price of the commodity plus a premium to cover losses during the month plus one month's interest equivalent to the rate of return to investment. The interest payment represents forgone earnings if the commodity was purchased with the storage agent's own financial capital; if the money was borrowed, at least part of the rate of return payment represents a financial obligation to pay interest. In both instances, payment for the return to storage is part of the capital cost.

Budgets that include storage costs must therefore match costs with the marketing margin in a temporal sense. This task usually requires a detailed understanding of market behavior. If available data describe annual average margins, for example, the analyst might be tempted to estimate costs of storage on a six-month basis. But, as Figure 10.1 shows, this calculation could be incorrect. The diagram traces commodity price behavior over time for rural producing areas, urban consuming areas, and imports. The world price shows no seasonality, being fixed at P_{cif}. In a global perspective, production of most commodities occurs at all times of the year and world prices show little seasonal behavior. Domestic postfarm costs for transportation and handling are represented by the difference between urban and rural market prices. Costs of storage are reflected by the rate of increase in the rural market price.

The key insight from the diagram is that domestic storage is not undertaken during all times of the year. In rural areas, prices decline from October through December (the harvest period) and the commodity is put into storage. Over the January through May period,

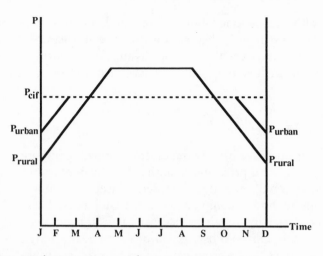

Figure 10.1. Marketing margins and storage costs

stocks are depleted and the commodity is taken out of storage to satisfy rural demands. But during the months of May through October, domestic stocks are eliminated and rural market prices no longer increase. Rural prices now reflect the cost of imports plus the cost of transport from the urban import point to the rural market. Storage is thus a six-month activity rather than a yearlong one, and the representative budget should be measured accordingly. An average storage period would be three months, January through March, and the return to storage would be represented by the difference between the March and January prices.

The recognition of intraseasonal price variation imposes a discipline on the output pricing for all the activities of the commodity system. In the current example, the farm-gate price is the January price (adjusted for transport costs between the farm gate and the rural market). The January price represents the purchase price of the commodity for the farm-to-processor activity. This item appears in the commodity-in-process category of the latter budget. The March price is the sales price for the farm-to-processor activity and thus appears in the output category of this budget.

If the costs of transport and handling are small or if the postfarm activities are unimportant to the policy issues for analysis, development of detailed activity budgets will be unnecessary. In this circumstance, the analyst can assume that costs are equal to observed marketing margins or can develop implicit margins based on price quotations from transport companies and rough estimates of storage costs. These private

costs are allocated among labor, capital, land, and tradable-input categories. Each cost item is then revalued in social prices. Because these
activities reflect services that are not available on world markets, the
social price of each activity will necessarily equal the sum of the social
costs.

Processing Budgets

Processing costs are usually the most prominent component of postfarm costs. Detailed price and quantity data for inputs and outputs of
processing thus will generally be needed. The analyst can visit representative firms to find a comprehensive list of inputs. Labor, fuel and
lubricants, electricity, repair services, and packaging materials are variable inputs common to most agricultural processing activities; chemicals and additives can be prominent for some products as well. For most
processing technologies (household processing is the principal exception), fixed costs are a large share of processing costs. Processing machinery, buildings and storage facilities, and equipment to handle and
transport both raw materials and the processed commodity often dominate other categories of input costs. Hence, particular attention must be
given to careful description of fixed input requirements.

Since the annualized costs of fixed inputs have to be converted into
cost per unit of output, detailed knowledge of annual throughput or
capacity utilization is needed. Expected annual throughput is the preferred measure by which to determine annual cost, because this value
represents the level of demand that motivates the long-run decision of
the firm to operate. The expected value of throughput may differ from
its observed value in a particular year: firms might be observed in a
start-up stage, where construction of facilities is based on expected
future increases in demand; infrastructural constraints, such as availability and reliability of power supply, storage capacity, or availability
of transportation equipment, sometimes cause operating levels of firms
to be temporarily low; and business cycles cause aggregate demand to
fluctuate from year to year, particularly if the product has an income-
elastic demand. Visits to a large number of firms and time-series data
can be helpful in the formation of an estimate of expected annual
throughput. When capacity utilization rates vary widely and chronically, multiple processing budgets can be constructed to represent alternative scenarios.

The valuation of outputs and the selection of outturn ratios are often
interdependent, because the quality of output and price are inversely
related to the outturn ratios. With rice or wheat, for example, process-

ing yields can vary substantially within a single processing technology. Outturn ratios can be higher for parboiled brown rice than for white rice and higher for whole wheat than for white flour. But the prices of the alternative outputs differ as well. If they did not, processors would increase profits by producing the product with the highest outturn ratio. In addition, outturn ratios influence the quantity and quality of by-products, such as the bran collected from cereal processing. These by-products often are of significant value relative to processing costs. Rice, for example, is frequently milled for a nominal or zero charge in exchange for the bran; cotton ginning can be done similarly, with cotton-seeds the relevant by-product.

Survey Strategies

Rarely are postfarm activities given the degree of attention accorded the farm sector. Few developing countries have transportation ministries or ministries concerned exclusively with agricultural processing. Secondary data for costs and returns are thus scarce. The analyst will be forced to visit a wide range of government ministries and academic/research institutions to seek out specialists with information about postfarm activities. Central statistical agencies may carry out annual or occasional censuses of manufacturing that include firms processing agricultural commodities. Ministries of agriculture may devote some attention to postfarm activities of the most important commodities. But in most cases, secondary data will not be sufficiently detailed to allow the construction of budgets.

Fortunately, limited primary surveys are usually feasible. Firms engaged in postfarm activities are much fewer in number than are farmers. Marketing firms are easily located (a necessary feature of their business), and larger firms often maintain written records of expenditures and receipts. Moreover, the inputs used in most postfarm activities (except processing) are relatively small in number. Survey efforts can focus on the activities that are most important in total postfarm costs and can summarize the less important activities with budgets of only two or three items.

Surveys of Small-Scale Firms

Direct field interviews are the most common way to compile budget data for small-scale transport or processing activities. Written records of inputs and outputs exist only rarely. Small-scale firms are usually owner operated and often utilize makeshift capital equipment. If these items are not available on markets, their current market-equivalent value—the amount the activities would pay if the input were to be purchased—needs to be determined. If capital equipment items (such as the principal machinery) are purchased, the activity operator might not be aware of current prices but might know the useful life and salvage values. Current prices must then be found through equipment suppliers or processors and marketing agents who have recently built or retooled their operations. New businesses can also be good sources of information about replacement costs for buildings and other infrastructural capital. If current prices cannot be obtained, last-resort approximations are generated by application of an index of inflation (producer prices or wholesale prices) to historical purchase prices.

Data collection for variable input use requires several strategies, particularly when written records are not kept. Because annual use of inputs such as fuel, lubricants, and labor usually cannot be recalled with accuracy, such data are most easily collected on a per day basis, a per hour operation of machinery basis, or a per unit of output basis. Eventually, all information must be converted into a common numeraire and a common time frame. But primary data collection should include whatever numeraires and time frames yield the most accurate responses. In each case, conversion ratios will be necessary to convert costs and returns to an annual or per unit output basis.

The final category of data involves outputs. Annual throughput might be well known by the activity operator, and this information is essential to determine fixed input costs per unit of output. Alternatively, the activity operator can be asked to describe output activity during an average week or month; in this circumstance, information about the number of weeks or months of operation per year is necessary as well. Actual or imputed market values of secondary outputs must be included along with primary outputs. In many small firms, output prices will be unavailable, because the activity operates largely on a custom basis and neither buys nor sells the commodity. This practice is recognized in the structure of the activity budgets by the assumption that the processor buys the commodity from the farm-to-processor activity. The custom processing fee is then added to the simulated commodity purchase price to obtain the simulated sales price for the output of the processing activity.

Surveys of Large-Scale Firms

Investigations of large-scale marketing operations amount to industrial firm surveys. These investigations almost always focus on processing. Because record keeping is quite detailed, surveys usually can obtain precise estimates of costs and returns. Some of the data necessary for the development of representative budgets can be gathered from annual financial reports. These statements provide measures of outputs, stocks, and raw material (commodity-in-process) inputs. But other inputs will not be described in sufficient detail for the PAM. Even reported capital equipment values may be of limited use. Accounting statements usually reflect some depreciated value based on the historical purchase price of the fixed input, and this reported value might be an unreliable indicator of replacement costs. Tax laws also can alter the ways that depreciation and book values are reported. Data from the country's industrial census might be helpful if they are reliable and up to date, but direct interviews and questionnaires are generally needed.

Because industrial technologies are highly complex, site visits and trials of test questionnaires will be needed to develop a clear under-standing of input-output relationships. The actual survey can be com-pleted in a relatively brief time period, because the number of large-scale firms is usually small. Screening firms during the initial site visits can save time for the research effort by allowing a focus on representative firms that have accessible and high-quality information. If the analyst is concerned about bias in results, a random sample can be chosen. But differences in responsiveness mean that the budget constructed will be based ultimately on only a subset of this sample. Box 10.2 illustrates a typical questionnaire format used for a study of wheat flour mills in Portugal.

A particular base year is desired for the study, although some flex-ibility might be needed to accommodate the accounting procedures of firms. If the study is performed on a calendar year basis but the account-ing year involves some different twelve-month period, data are re-quested for the year closest to the base year. When annual inflation rates are very high, budgets based on accounting year data must be adjusted. But in most cases, the accounting year estimates will not need to be changed much in order to approximate results for the calendar year.

Guarantees of confidentiality are also important for successful data collection. When asked for details about costs, revenues, and profits, firms are understandably sensitive about how that information will be used. But because activity budgets are built for representative firms, no individual firm's practices will be revealed in published results. Discus-sions with top-level management and presentation of copies of research project proposals during initial site visits enhance effective information collection.

Once the data for inputs and outputs are collected, the questionnaires are combined to generate a synthetic representative budget. Because

Box 10.2. Sample Questionnaire Topics for Large-Scale Processing Mills

A. General Information
    1. Name and address of firm.
    2. Year production began.
    3. Dates of accounting year.
B. Production and Sales
    For each product:
    1. Opening stocks (quantity and value).
    2. Production (quantity and value).
    3. Sales (quantity and value).
    4. Closing stocks (quantity and value).
C. Capital Stock
    Construction or purchase date, initial costs, current value, salvage value, expected useful
    life, and estimated replacement cost for each of the following categories:
    1. Plant.
    2. Machinery.
    3. Residential buildings.
    4. Furnishings.
    5. Tools.
    6. Transportation equipment.
    7. Other (specify).
D. Labor Force
    Number of full-time employees (in full-time equivalents), cash wages and salaries, value of
    fringe benefits, employer's taxes on labor (if any; specify) for the following categories:
    1. Managers.
    2. Supervisors.
    3. Skilled laborers.
    4. Unskilled laborers.
E. Intermediate Inputs
    List, by item, the quantity and cost of all material inputs used in production, as well as
    parts, containers, nondurable tools, and spares—that is, all nonlabor and nonfixed inputs
    except for raw materials.
F. Raw Materials (Commodities in Process)
    For each commodity:
    1. Opening stocks (quantity and value).
    2. Purchases (quantity and value).
    3. Closing stocks (quantity and value).
G. Land Area and Estimated Current Market Value
H. Capacity Utilization
    1. Number of days of operation in year.
    2. Length of single shift, in hours.
    3. Number of shifts of operation in year.
    4. Estimate of full one-shift capacity, in raw material input per hour and potential
       number of days of operation per year.
    5. Estimate of full two-shift capacity, in raw material input per hour and potential
       number of days of operation per year.
    6. Estimate of full three-shift capacity, in raw material input per hour and potential
       number of days of operation per year.

firms differ in capacity and level of annual throughput, no two firms will report identical numbers for input use and output. The first task for budget preparation thus is to group the firms into size categories, according to processing technology. The measures for size categories are usually related to hourly processing capacity, such as number of metric tons per hour. Choices about the range of capacities to include within a single group depend on the presence of economies of scale. If a flour mill with a 4 metric ton per hour capacity is roughly the same as two 2 metric ton per hour capacity mills, both types of firms can be grouped together.

The choice of representative capacity is made in concert with the valuation of fixed capital costs. Capacity can be an average of the capacities of the sampled firms, but the availability of information on fixed investment costs often requires more pragmatic choices. Because firms within a grouping will have plants of different ages, replacement costs for fixed inputs might not be estimable from questionnaire results. The analyst will then be forced to use capital stock evaluations from the more recently established firms (adjusted for inflation, if necessary). Discussions with engineering and construction firms can also be helpful to verify the representativeness of particular fixed capital cost estimates.

Evaluations of labor, intermediate input, and raw material input requirements can be determined by comparison of the responses across firms within the sample. The input quantities and values can be standardized by expression of the reported results per unit of raw material input or per unit of processed product output. When these variable inputs are used in constant proportions, the standardized estimates should be comparable across firms in the sample. But if the standardized estimates are very different, some causal explanation must be sought. The sample is reclassified and multiple budgets are developed to reflect differences among firms in production technology.

Differences in shift work across firms often are important to the size and cost of the labor force. Administrators and many skilled employees often do not work during second- and third-shift operation. High wages and overtime payments might be required for those who work second and third shifts and for weekend and holiday work. The analyst has to choose between simple averages of results across firms and specific responses in shaping the budget of the representative firm. In either case, choices about shift work must be consistent with the selected rate of operation.

Partial Budgeting

Once budgets are constructed, partial budgeting can be used to evaluate the impact of additions to or substitutions of equipment in the existing technology and alterations in the rate of capacity utilization of equipment. Additions and substitutions may be constrained by the technical compatibility of components. Often, technological change in processing industries requires complete replacement of existing equipment. But in some cases, particularly with storage and handling facilities, components of the processing firm may be operationally separable from one another. Interviews with engineers and processing technicians can be used to verify the viability of new budgets.

Changes in capacity utilization may be feasible for several reasons. Monopoly power among processors, government-induced distortions in hours of operation, and the costs of capital equipment can create conditions where operation times are well below technically feasible levels. Capacity utilization can also be influenced by the costs of collecting raw materials and distributing the processed product. A reduction in these costs, either by elimination of distortions or by introduction of a new infrastructure, may allow processing firms to expand their markets. Because fixed costs are prominent in total costs of processing, reduced utilization rates can cause observed processing margins to be substantially larger than their cost-minimizing levels. It is also possible that entirely different processing technologies would be chosen under more intensive utilization. If the analyst is examining a sufficient number of representative processing systems, this possibility can be evaluated by comparison of the costs of alternative techniques, each operating at full capacity utilization.

The maximum feasible capacity utilization rate will almost always be less than 100 percent. Equipment downtime is often mandated by maintenance and repair requirements, social constraints on operating at certain periods, or limitations of local market demand. Interviews with technicians and other industry experts are necessary before the maximum capacity utilization rate can be chosen. Introduction of an adjusted capacity utilization rate will influence the fixed input costs of the processing budget. Some variable input adjustments might be required as well; as mentioned, higher wages usually are required for night and weekend operation. But most variable inputs will vary in direct proportion to output. Because the budgets are already calculated in terms of costs and revenues per unit of output (or per unit of raw material input), the greater part of the original representative budget can be directly transferred to the budget that simulates costs and returns under higher rates of capacity utilization. Box 10.3 describes a partial budgeting analysis of full capacity utilization for flour mills in Portugal.

Concluding Comments

The procedures for compiling budgets for representative postfarm activities are very similar to those used in the compilation of farm budgets. Descriptions of marketing chains are analogous to descriptions of cropping calendars, and input and output valuations confront the same problems of separating prices and quantities. Two principal differences distinguish postfarm from farm budgets: primary surveys are easier to do off the farm than on it, and the relative prominence of fixed costs means that economies of size usually are more important for postfarm activities.

Research projects that estimate PAMs in practice often give limited attention to postfarm activities. In particular, little research effort is expended on the disaggregation of costs among inputs. Such treatment is justified only when postfarm costs are trivial or when the farm commodities can be directly marketed in world markets. In other circumstances, postfarm costs are an integral part of the commodity system. Analysis is needed to assess the impacts of policies and market failures on the postfarm activity. These divergences may be as important as or more important than those influencing the farm activity in a commodity system. Even when the primary concerns are with farm-level issues, postfarm divergences can have major implications for farm prices and incomes.

Bibliographical Note to Chapter 10

Literature concerning the conduct of marketing and processing surveys is even less common than farm survey literature. On agroindustrial activities and their linkages to farm activity generally, see James E. Austin, *Agroindustrial Project Analysis* (Baltimore: Johns Hopkins University Press, 1981). J. Price Gittinger, *Economic Analysis of Agricultural Projects,* 2d ed. (Baltimore: Johns Hopkins University Press, 1982), chap. 5, provides a detailed discussion of the use of accounting data to evaluate large-scale processing firms. The estimation of transportation project costs is discussed in Hans A. Adler, *Economic Appraisal of Transport Projects: A Manual with Case Studies* (Baltimore: Johns Hopkins University Press, 1987). Discussion germane to both marketing and processing surveys is provided in Walter P. Falcon, William O. Jones, and Scott R. Pearson, eds., *The Cassava Economy of Java* (Stanford, Calif.: Stanford University Press, 1984).

One of the earliest works on developing country protection of agroindustrial

Box 10.3. Partial Budget Adjustments for Utilization of Full Capacity of Large-Scale Flour Mill in Portugal

The base-case budget for the representative large-scale wheat flour mill utilized an average operating rate of 240 days per year, eight hours per day. This rate was a single-shift operation, with annual throughput of 6,720 metric tons of wheat. Mill processing capacity was 3.5 metric tons of wheat per hour. The following table summarizes annual labor cost and total fixed cost estimates.

Input	Quantity	1981 value (thousands of escudos)
Labor force	1 administrator/manager	850
	3 engineers	1,410
	14 unskilled laborers	4,760
Capital equipment	Buildings	67,750
	Machinery	57,850
	Land	30,000

Discussions with plant managers allowed the development of estimates of potential maximum operating times. About 36 days per year (3 days per month) were needed to clean, repair, and adjust equipment; holiday regulations required the firms to be closed for 10 days. Therefore, maximum capacity operation was set at three shifts per day, 320 days per year (250 weekdays, 50 Saturdays, and 20 Sundays). Annual throughput increased to 26,880 metric tons of wheat. Although no premia were paid for shift work, Saturday labor commanded a 75 percent premium above regular wages and Sunday labor a 200 percent premium. The skilled labor force required only a small increase in size. The following table summarizes annual labor cost and total fixed cost estimates for full capacity utilization.

and other industries is Stephen R. Lewis and Stephen E. Guisinger, "Measuring Protection in a Developing Country: The Case of Pakistan," *Journal of Political Economy* 76 (December 1968): 1170–98. A number of studies were conducted in the 1970s; perhaps the broadest group is the twelve-volume set (including ten country studies) by Jagdish Bhagwati and Anne O. Krueger, eds., *Foreign Trade Regimes and Economic Development* (New York: National Bureau of Economic Research, 1975).

The presence of economies of size in processing has been long recognized, but discussions of the relationships between capacity utilization and distortions are relatively recent. A seminal theoretical discussion is provided in Gordon C. Winston and Thomas McCoy, "Investment and the Optimal Idleness of Capital, *Review of Economic Studies* 41 (July 1974): 419–28. Empirical applications of social profitability analysis to economies of size in agroindustries are provided in William F. Steel, "Import Substitution and Excess Capacity in

Input	Quantity	1981 value (in thousands of escudos)
Labor force	1 administrator/manager	850
	4 engineers	1,880
	42 unskilled laborers	24,419
Capital equipment	Buildings	67,750
	Machinery	57,850
	Land	30,000

Further investigation revealed a potential technical change—the addition of a flour silo and automated sacking equipment—that had been explicitly prevented by regulation. These investments had been made by one firm in anticipation of a change in the regulations, allowing simulation of the new technology. Full capacity costs are presented in the following table.

Input	Quantity	1981 value (in thousands of escudos)
Labor force	1 administrator	850
	5 engineers	2,350
	18 unskilled laborers	10,465
Capital equipment	Buildings	85,143
	Machinery	64,061
	Land	30,000

The direct labor costs were converted to a per metric ton of flour basis and substituted in the base-case activity budget. The fixed costs were converted to annual equivalent values, adjusted to per metric ton of flour basis, and substituted in the base-case activity budget.

Ghana," *Oxford Economic Papers* 24 (July 1972): 212–40; Scott R. Pearson and William Ingram, "Economies of Scale, Domestic Divergences, and Potential Gains from Economic Integration in Ghana and the Ivory Coast," *Journal of Political Economy* 88 (October 1980): 994–1008; and Eric A. Monke, Scott R. Pearson, and J. P. Silva-Carvalho, "Welfare Effects of a Processing Cartel: Flour Milling in Portugal," *Economic Development and Cultural Change* 35 (January 1987): 393–407. A further illustration of the use of the PAM approach in studying agroindustries is provided in Michael Barzelay and Scott R. Pearson, "The Efficiency of Producing Alcohol for Energy in Brazil," *Economic Development and Cultural Change* 30 (October 1982): 131–44.

CHAPTER 11

Estimating Social Profitability

THIS CHAPTER is concerned with the empirical estimation of social costs and returns, the most complicated analytical task in the construction of the PAM. The information requirements for exact calculation of social prices of outputs and inputs are so vast that empirical estimates will never be exact, even for the best-understood economies. The policy analyst is concerned with whether approximations of social costs and returns will be sufficiently close to their true values to allow useful insights into the motivation for existing policies and the potential gains from changes in policies.

For tradable inputs and outputs, social valuation entails calculation of world price equivalents for the domestic product and requires particular attention to the effects of variations in quality and in geographic location. For domestic factors, the social valuation process begins with observed market prices and then adjusts those prices for the effects of factor market divergences. Estimates of the other influences on factor prices—the interactions between output divergences and factor prices and between input substitution and factor prices—are often no more than informed guesses. These effects usually are assumed to be small in magnitude. Because social factor price estimates necessarily are approximations of true values, sensitivity analysis of the effects of changes in social factor prices is a key element of the presentation of PAM results. By altering the quantities of inputs and outputs from the values observed under private price incentives, the analyst can also simulate producer response (if any) to the social prices for outputs and inputs.

Estimating Social Prices for Tradables

The social price for an agricultural commodity is a border price—the price at which foreign suppliers would deliver the commodity to the domestic market or the price that foreign consumers would pay domestic suppliers to deliver the commodity to their markets. But if the world price is to represent an expected social opportunity cost for the domestically produced commodity, it may require adjustment. Observed world prices may not reflect the impacts of the domestic country's market power. In the absence of actual imports or exports of the domestically produced commodity, world price equivalents must be estimated. These prices will usually be derived from some observed world price. But to be truly equivalent, they must reflect the effects of international transport costs and differences in quality. Moreover, observed world prices may be different from those expected in the future.

Finding World Prices

One source of data on world prices is the country's international trade statistics. Implicit world prices (average per unit values) can be found by division of total values by quantities traded (of imports or exports). But biases in trade data can arise if the amounts traded are small, if the data are distorted because firms have improperly reported figures to avoid taxes or to repatriate earnings (overinvoicing or underinvoicing), or if the quality of the foreign produced commodity differs substantially from that of the domestically produced commodity. Interviews with officials from the statistics agency or from trading companies allow assessment of these possibilities. If own-country trade data are missing or biased, comparable world prices often can be estimated by examination of trade data of a nearby country. These prices are adjusted for international transport and insurance costs to simulate a cif import price or fob export price for the country of interest. A third possible source of direct world prices is price information from industry, government agencies, or international organizations. These groups regularly publish cif or fob prices at various locations, which can then be adjusted to allow for any international transport and insurance cost differences between the listed port and the relevant country port.

When direct world price information cannot be found, an alternative procedure is to estimate world prices indirectly, by removing the effects of distorting domestic policies. To find world prices indirectly, one starts with the private prices of tradables and then estimates the quantitative impact of policies affecting the commodity. This process results in an implicit world price that reflects the removal of the distorting effects of policy. In terms of the PAM, I or J is measured in order to find E or F as residuals from A or B, respectively. The procedure works well when all policy transfers are easily measurable and known. If, for example, the sole distorting policy on an importable fertilizer input is a subsidy that is known to be 20 percent of the cif import price of fertilizer, the observed domestic price of fertilizer is 80 percent of the world price (calculated as the domestic price divided by 0.8).

This procedure is more difficult when quantitative trade restrictions are present or when transfers involve a combination of policies, including quotas. For instance, rice imports might face a 30 percent tariff and a quota of 500,000 metric tons. In adjusting an observed domestic price of 450 rupiah per metric ton, the researcher cannot simply divide 450 by 1.3 and claim the implied world price (cif local ports) is 346 rupiah. The quota on rice imports, if binding, would create upward pressure on domestic prices in addition to that already caused by the tariff. The implied result, 346 rupiah, sets an upper limit to the world price, but the actual world price could be much lower. In this circumstance, cif or fob rice prices in nearby countries must be obtained.

Location

Recognition of positive costs for transportation and handling is one potential complication in the identification of world prices. The price received by domestic producers for exportation to foreign markets (the fob price) is no longer the same as the price paid by domestic consumers for importation from foreign markets (the cif price). The foreign demand schedule becomes distinct from the foreign supply schedule, as shown in Figure 11.1. If the country is a net importer of the commodity, domestic demand and supply relationships are like those illustrated by the supply curve, S, and the demand curve, D. The domestic market price under free trade will be P_M^W; this term represents the social value of the commodity. If the country is a net exporter of the commodity, the domestic market supply curve will be represented by S'', and the social value will become a lesser amount, P_X^W. If the supply curve intersects the demand curve at a price between P_M^W and P_X^W, as does S', no trade occurs; the social price of the nontradable commodity, P_D, is determined entirely by domestic market conditions.

When international transportation and handling costs are small, the

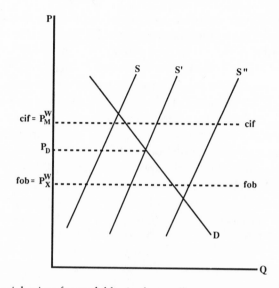

Figure 11.1. Social prices for tradables in the small country case

cif-fob price band will be small as well. Whether fob or cif prices are used to represent social values is a minor concern. But when the price band is large, the choice of appropriate world prices becomes an issue for the calculation of social profitability. The problem for construction of PAMs arises because the location of the supply and demand curves under social price conditions is not known. The observed supply curve reflects the impacts of divergences in the markets of related commodities and markets for domestic factors and commodity inputs. A commodity that is imported or nontraded might be an exportable under social price conditions. Hence, judgments about quantities of domestic supply and demand under social price conditions become necessary elements in the choice of social output prices.

The possible wide differences between social profitability calculated at cif versus fob prices means that location can be a key criterion in the identification of commodity systems. Specification of a system as producing for the export market allows direct evaluation of competitiveness at the fob price. Social prices for internal domestic market locations are then estimated by subtraction of the social costs of transport and handling from the fob price. This procedure generates a schedule of social prices for the exportable. If internal prices were to fall below this fob-determined price, producers would increase exports, causing internal prices to rise; prices above the fob-determined price would cause a diversion from exports to internal markets, forcing internal prices to fall. If, instead, production is substituted for imports, cif prices are the relevant border prices. Domestic transport and handling charges are then added to the cif price to estimate social prices for internal market locations. Arbitrage (between importing and consumption of domestic production) again assures that this set of social prices will be sustained.

When the commodity is nontradable, the social supply and demand curves intersect within the fob-cif band. In these circumstances, only upper and lower limits can be defined for the social prices. The actual price is determined by local supply and demand. Because the social curves are not known, the social output price bears some degree of indeterminacy. These indeterminacies are likely to be most important at isolated markets, because differences between the cif-determined and the fob-determined price schedules become increasingly large as internal markets are more distant or more isolated from the border.

Time Frame of Analysis

Explicit recognition of the time frame of analysis is the second aspect of identification of social commodity prices. Observed domestic market prices and world prices can be used to assess the profitability of the commodity system in any given year. From the policy-maker's perspective, however, long-run profitability is often more important. To capture longer-run interactions of policy and profitability, the analyst needs to use expected prices as the measures for calculation of input costs and output revenues. Generally, observed prices will be close or equal to expected prices, and the distinction between expected and actual prices will become a nonissue. Because most commodity world markets are large, long-run demand and supply shifts in world markets generate slow changes in commodity prices. But if the analysis is conducted in an unusual year, when extraordinary demand or supply shifts create abnormally high or low prices, the use of observed prices can give misleading measures of profitability and policy transfers.

Estimates of expected future prices can be obtained from commodity models of world markets. Price projections are made for a number of commodities by international organizations such as the World Bank. In principle, a price projection represents the current price adjusted for expected future changes in the demand and supply balance. If demand is expected to increase more rapidly than supply, expected future prices will be higher than current prices; in the opposite circumstance, expected future prices will decline. The difficulty with such exercises is the possibility that they will be self-defeating prophesies. If market participants believe the projection and thus expect future prices to rise, incentives are created to find new sources of supply or substitutes in consumption. If prices are expected to decline, producers will be encouraged to switch to other activities and new sources of consumer demand will be encouraged. In part because of such adjustments, eventual market prices usually differ substantially from their projected values.

Because of such complications, the current price is usually the best indicator of expected prices in the near future, unless it has been influenced by some large and unusual shock in the world demand-supply balance. The rationale for reliance on current prices is particularly strong when the commodity can be stored, because expected future prices will be linked to current prices. If expected future prices exceed current prices by more than the cost of storing the commodity, an incentive will exist for merchants to buy the commodity at the current time and store it for future sale. This action will increase current prices. If expected future prices differ from current prices by less than the cost of storage, incentives will exist to sell some of the stored commodity in the current period, thus driving down current prices. In either instance, intertemporal arbitrage should equate expected future prices to current prices plus storage costs.

In many agricultural commodity markets, world prices are influenced by subsidy and trade policies in foreign countries. Even if all countries in the world market were price-takers, the aggregate impact on world price could still be significant because so many countries use these policies. From the efficiency perspective, however, the magnitude of this price effect is not an important issue. For an individual country, optimal income is generated by using world prices, however imperfectly they might be established. The world price thus represents the opportunity cost of the commodity for the domestic economy. Domestic policymakers might feel that foreign country policies are unfair to domestic producers; in this case, a decision to set a private market price greater than the social (world) price could use foreign policy considerations as a nonefficiency rationale. The particular price effects of foreign subsidy and trade policies thus become important only if future policy changes are expected to alter world prices. For example, reductions in subsidies in exporting countries or reductions in tariffs in importing countries would cause world prices to rise. Because prediction of the timing, magnitude, and price effects of such policy changes is usually impossible, most estimates of expected prices implicitly assume constancy of agricultural policies in other countries.

A final complication for expected price calculations arises because of world price variability and consequent risk premia. In some commodity markets, world prices fluctuate widely from year to year. Such conditions most often characterize residual markets, where world exports or imports (or both) are dominated by the policy actions of one or a few large countries. Relatively small changes in demand-supply balances in these countries can then translate into large changes in world market supply and demand. In response to such conditions, countries may try to pursue stable internal prices by varying positive subsidies to consumers of importables or to producers of exportables. When domestic production fluctuates, net demands from the world market or supplies to the world market are changed. If many countries pursue such stabilization policies simultaneously, and if net demand changes are correlated across countries (for example, when countries are subjected to similar weather), world price instability is further aggravated.

In the presence of substantial price variability, small countries may believe that long-run averages of world prices are an insufficient measure of the incentives afforded by world markets. If consumers are willing to pay a premium for price stability, the social value of output may be increased by a tax, the proceeds of which finance a domestic price stabilizing scheme. Such schemes can be pursued with a variety of methods (buffer stocks, financial buffer funds) and provide varying degrees of food security, depending on the size of the buffer. The choice of risk premium is usually based on the net cost of operating an arbitrarily specified buffer program.

Quality

Quality differences between the product in the domestic market and the competitor available from the world market constitute a third complication in the calculation of social commodity prices. Premia and discounts can usually be associated with particular characteristics, such as appearance, impurities, nutrients, and country of origin. If the product is an export, the export price data already reflect the quality of the domestic product. To find implicit world prices for importables, however, judgments are needed to establish comparable quality so that observed world market prices can be adjusted to match the quality of domestic output. The relative domestic prices of the imported commodity and its local counterpart reflect the marginal valuations of domestic consumers. Because preferences differ among countries, the world market may place a different relative price on the domestic commodity. But if many commodities of different quality are traded on

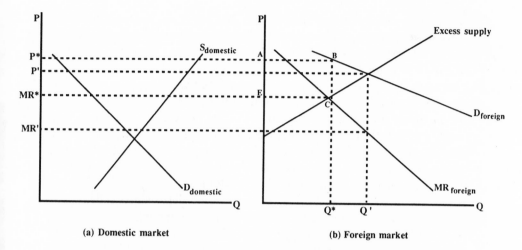

Figure 11.2. Social prices for outputs under large country assumptions

the world market, the empirical task involves specification of the physical characteristics of the domestic commodity and the selection of a commodity of comparable quality from those available on the world market.

Large Country Effects

When a large country can influence world prices, the selection of appropriate social prices depends on domestic supply behavior and the characteristics of world market demand. Figure 11.2 considers the case of a large country exporter. Figure 11.2a describes the domestic market and Figure 11.2b the foreign market. Because foreign market demand is downward sloping, the foreign marginal revenue curve must be downward sloping as well. Domestic supply to the foreign market is derived by estimation of the difference between domestic supply and demand at different prices. Under competition, these supply prices are the same as marginal costs of production. Maximum profits are achieved when foreign marginal revenue equals domestic marginal costs; this point occurs when exports are Q^*. To hold exports to this level, the government imposes an export tax equal to $P^* - MR^*$. Export tax revenues amount to $(P^* - MR^*)Q^*$, or ABCE on the right-hand side of the figure. Under export taxation, the price received by producers is MR^*, not P^*. If the export tax is not used, competition among domestic producers will lead to an output of Q' and a price of P'. In this situation, income to the economy is less than it might be, because marginal production costs (equal to P') are greater than marginal revenue, MR'. But unless producers receive the export tax revenue, they will be better off with unregulated trade.

In principle, tax-inclusive prices should be used for social valuation, because their use creates the maximum national income. In the presence of an optimal export tax, private and social prices will be equal, and both will be less than the world price; for an optimal import tax, private and social prices will be in excess of world prices. In practice, however, the detailed empirical models of commodity market behavior necessary to permit the calculation of world and domestic prices under optimal taxes are often lacking. In part, the lack of models reflects a dearth of data. But for many agricultural commodities, judgments about market power are also confounded by the prominence of policy interventions. Particular countries can have a large share of world trade. If their share of world production is small, however, attempts to impose optimal taxes may result only in policy changes within competing nations that cause the tax-imposing country to lose market share rather than influence world prices. In most empirical situations, therefore, optimal tax arguments can be raised as a justification for an observed divergence between private prices and world prices. But such measures are not likely to indicate the optimal size of the divergence.

Using the Exchange Rate

At some point, the analyst needs to convert world prices into domestic currency; this conversion requires an exchange rate. Entries in the calculation of private profitability, the top row of the matrix, present no difficulties. The researcher normally does not even come across an exchange rate in private budgeting. If some items are denominated in foreign currency, the actual exchange rate used by the farmer or marketing agent—official or otherwise—is the correct candidate for conversion. Private profits measure observed incentives and results. When policy is not enforceable—for example, when parallel market rather than official exchange rates are used by participants in the system—there is little point in assuming that private actions are affected by nonbinding policy. Private profits are measures from actual markets, whether legal or illegal.

Interest then centers on what exchange rate to use to convert world prices into domestic currency for social valuation (in the second row of PAM). Adjusting the exchange rate for the impacts of output price distortions and macroeconomic policy effects is a complex task. Fortunately, such corrections are not essential for the construction of the PAM. Exchange-rate changes cause changes in output prices that will be transmitted eventually to domestic factor prices. Social factor prices reflect marginal value products—the social price of output times the marginal physical products—and these prices will change in equal proportion because changes in exchange rates alter tradable-output prices in equal proportion. All tradable-commodity systems are similarly affected by the exchange rate, once factor prices have had time to adjust. The nonuniform effects of exchange-rate changes occur in systems for nontraded goods, because the nontraded-output prices are not directly affected, whereas all input costs are changed. Producers of nontraded commodities will face pressures to alter input combinations and reduce costs; otherwise, they will be forced to raise output prices and suffer the consequences of reduced demand.

The inputs and outputs of tradable-commodity systems are not identically affected, however, when the government fails to offset inflation with exchange-rate changes. In this circumstance, factor prices increase faster than the exchange rate depreciates and thus faster than prices of tradable outputs and inputs rise. PAM analysts then need to adjust private exchange rates (and the social prices of outputs and tradable inputs). This inflation adjustment simply corrects for past or projected movements in the country's real exchange rate (RER). The RER is found by comparison of the country's ratio of exchange-rate changes to wholesale (or consumer) price changes with the same ratios for its major trading partners.

An example serves to demonstrate the process of RER adjustments. If a country experiences an inflation rate of 50 percent while average inflation in the country's major trading partners is only 5 percent, differential inflation is about 43 percent: $(150 / 105 - 1) \times 100$ percent. The country's exchange rate is pegged to the U.S. dollar and is initially (before the differential inflation) in equilibrium at 100 domestic currency units (DCUs) per U.S. dollar, or DCU 100 / $1. If the government maintains the fixed exchange rate, the real exchange rate will appreciate to DCU 70 / $1 (the end-of-year official exchange rate, DCU 100 / $1, divided by a ratio of the indices of the domestic inflation rate to the weighted average inflation rate in the principal trading partner countries, 150 / 105). If this year is used as the base year for PAM analysis, an exchange-rate correction of 43 percent will be required to estimate the equilibrium real exchange rate prevailing at the end of the year. The inflation adjustment factor will be 1.43, and the adjusted end-of-year rate will thus be DCU

143 / $1. If differential inflation is assumed to occur uniformly through-out the year, the average degree of overvaluation will be half of 43 percent and the average adjusted exchange rate will be about DCU 122 / $1. The social domestic currency prices of tradable outputs and inputs will then be 22 percent higher than private prices (122 / 100), reflecting the inflation tax that distorting exchange-rate policy imposes on tradables.

A second circumstance in which the social prices of tradable com-modities might be adjusted for exchange-rate effects occurs in the pres-ence of multiple exchange-rate regimes. These regimes involve a set of official rates or, more commonly, an official rate and a parallel market rate. The exchange rate selected to convert tradable-commodity prices should match the rate that is determining the observed level of domestic factor prices. When the official rate is controlled and parallel markets for foreign exchange are small, the official rate provides the relevant standard. When parallel markets predominate, the official rate should be ignored in favor of the parallel market rate. Finally, when parallel and official markets are of comparable magnitudes, a blend of the two rates is appropriate. In all cases, the analyst must discover the particular rate used to generate the domestic prices for the tradables of the com-modity system. For example, if trade data used for social values are based on the official exchange rate but parallel markets dominate in the economy, the social values of tradables (items E and F of the PAM) are multiplied by the ratio of the parallel rate to the official rate.

Finally, exchange-rate adjustments might also be desired to reflect certain short-run effects of exchange-rate changes. Prices of tradable commodities are usually thought to adjust more rapidly than domestic factor prices, thus altering profitability. In the PAM, short-run effects of the exchange rate involve adjustments to E and F; because G does not change, H changes to a greater extent than the exchange rate. These calculations can indicate incentives for existing systems to change and for new systems (different technologies or different regions) to initiate production.

An enormous amount of empirical detail is required to calculate the social exchange rate—the magnitude of the sustainable government budget deficit, the extent to which the existing deficit is financed by foreign capital inflows, the slopes of the supply and demand curves for foreign exchange, and the magnitude of protection afforded various commodities. As a result, estimations usually entail much guesswork. Other experts on the domestic economy may provide useful reviews of assumptions and calculations. Domestic government organizations, such as the ministry of finance or the central bank, may have economic advisers who assess the economy at an aggregate level. These individuals should have information about the inflation rates and the structure of official and parallel market exchange rates. International organizations such as the World Bank and the International Monetary Fund may also have expert advisers doing such calculations. Box 11.1 provides a stylized example of the procedures needed to determine the social price of a tradable good.

Estimating Social Prices for Domestic Factors

Given sufficient information about world prices for tradables and technologies of production, social prices for factors could be calculated by direct evaluation of the marginal physical products at world prices. But without such information, less direct evaluation strategies must be pursued. The alternative strategy begins with observed private prices for domestic factors and adjusts those prices to take account of the effects of divergences.

In practice, the principal adjustments to domestic factor prices are corrections for the effects of factor market divergences. The remaining sources of change—output price distortions and input substitution effects—are usually assumed to be small enough to ignore. Ignoring output price effects is reasonable only if the pattern of output price distortions does not demonstrate a bias toward particular factors. If protection is given to commodity systems that span the range of factor intensities, the impact of one output distortion on factor prices (and factor use) will be offset by the countervailing impact of another output distortion. The aggregate impact of all output price distortions will then be small.

If output prices are systematically biased to favor a particular factor, factor price effects are usually assumed to be slight magnifications of the net bias in output protection. The zero-profit condition shows that the proportional changes in factor costs, each one weighted by its share in total cost, must equal the proportional changes in price. Therefore, the proportional factor price changes will magnify the net protection provided to outputs. If policy systematically protects capital-intensive commodities, for example, private rates of return to capital will be above social rates of return by an even larger proportion. Private prices for some other factor, such as labor, will be below their social values.

Justification for ignoring input substitution effects on factor prices

---

Box 11.1. What Is the Social Value of Rice?

The commodity system for analysis involves rice production in Mali, a landlocked country in West Africa. The problem for social valuation is to determine an appropriate social price for Malian rice. The analyst begins by assembling marketing year price data for various qualities of rice traded on world markets. The choice of a social price of rice depends on assessments of quality and the likelihood that recent prices will continue into the future.

Malian rice has a "brokens" content, on average, of about 35 percent. Thai Al super, a higher percentage broken variety, is selected as a variety that represents quality comparable to that of the Malian product. The next decision is whether the recent prices are representative of expected future prices. On the basis of a review of the market literature the analyst selects $175 per metric ton fob Bangkok as a measure of the long-run world price.

The next step in the valuation process is to analyze the costs of moving the rice from Bangkok to Bamako, the capital and main consumption center in Mali. This movement involves ocean transport from Bangkok to Abidjan, Ivory Coast, and then shipment by rail to Bamako. Because fob Bangkok and cif Rotterdam quotes are available, international transport cost margins can be estimated. These margins have ranged between $40 and $60 per metric ton in recent years, and the Bangkok-Rotterdam route is roughly comparable in costs to the Bangkok-Abidjan route. The analyst chooses $50 per metric ton as the ocean transport cost. Inquiries to Bamako importing firms reveal that shipping costs from Abidjan to Bamako are an additional $25 per metric ton. The long-run social price for rice cif Bamako is then estimated as $175 + $50 + $ 25, or $250 per metric ton.

---

arises from second-best situations and from the envelope theorem result: marginal responses to changes in costs of production are the same whether input-output coefficients are fixed or variable. But if factor price changes are very large, input substitution will have exerted some measurable impact on factor prices. Consideration of these more general circumstances requires information about the input demands of each commodity system of the economy; adequate data to measure the effects of import substitution do not exist in most empirical situations. In these circumstances, an analyst has little choice but to resort to sensitivity analysis to test whether the PAM results change much within ranges of assumed parameters.

Capital

Estimation of the social rate of return begins with observed interest rates in the capital market. The first adjustment to these rates involves a correction for inflation. If inflation rates are nonzero, savers and finan-

The next task involves conversion of the dollar price into Mali's local currency (the CFA). To illustrate the principles of calculation, a number of hypothetical and unrealistic assumptions are made in this example. The official exchange rate is assumed to be 700 francs per dollar at the beginning of the year. The government is assumed to depreciate the exchange rate by 10 percent, thus reaching 770 francs per dollar at the end of the year, or a midyear value of 735 francs per dollar. But if Malian inflation is 20 percent higher than foreign inflation, the franc should depreciate to 840 francs per dollar, or a midyear value of 770 francs per dollar. Hence, the inflation adjustment factor is 770/735, or 1.05. The midyear social value of foreign exchange is estimated as the official rate multiplied by the adjustment factor: 735 x 1.05 = 772. The social value for Malian rice in Bamako is thus $250 per metric ton x 772 francs per dollar = 193,000 francs per metric ton.

The value 193,000 francs per metric ton represents the social value of rice in the wholesale market. But the social value must be further adjusted for the social costs of internal transportation in order for social values for rice in the farm, farm-to-processor, and processor activities to be determined. It is assumed that all these other activities are located in the same area, 200 kilometers from Bamako. Transport costs are estimated at 15,000 francs per metric ton for that distance. The social price of rice for the processing activity becomes 178,000 francs per metric ton. Similar adjustments can be made to derive representative farm and farm-to-processor values (although these prices will need to be converted into paddy-equivalent prices). All of these prices reflect social values for a system that delivers rice to Bamako. If, instead, the competitiveness of rice production for consumption at the farm gate is to be analyzed, domestic transport costs will be added to, rather than subtracted from, cif Bamako values in order to generate estimates of social value.

cial intermediaries will require a financial premium on their savings and lendings, so that the real value of these transactions does not deteriorate over time. A consumer, for example, will not be much interested in saving if the savings will buy fewer commodities in the future than in the present. Observed (nominal) interest rates will thus reflect compensation for inflation as well as the real rate of return. The real rate of interest, i^B, is estimated as follows:

$$i^B = \frac{1 + i^N}{1 + f} - 1$$

where i^N is the observed interest rate and f is the inflation rate. At low rates of interest and inflation, the real rate of interest is well approximated by the simple difference of observed interest rates and the inflation rate, $i^N - f$. But this approximation worsens as the interest-rate levels increase, requiring use of the formula.

Real interest rates are then adjusted for divergences to derive real rates of return. The simplest divergence (for analytical purposes) is a tax on capital. If the tax can be represented as a simple proportional tax of rate t, the rate of return to investment (r) can be estimated from observed real interest rates with the following formula:

$$r - (r)(t) = i^B$$

The social and private rates of return must be higher than the real interest rate so that the investor can pay both the lender and the government.

Analogous calculations occur in the presence of proportional subsidies to capital investment $(- t)$. These policies are most commonly found as industry-specific credit or interest subsidy programs, and they affect only the private rate of return in subsidized systems. The interest-equivalent value of the subsidy (measured in percentage points) is subtracted from the economywide private rate of return in the estimation of the private cost of capital for the system. This procedure presumes that the potential borrower has access to investments that earn the economywide private rate of return and can use unsubsidized credit (not necessarily from banks) to pursue these investments. The investor will use subsidized credit for government-specified investments only if the net return from subsidized credit use equals the private marginal rate of return from unsubsidized credit use.

Typically, divergences in the capital market will be far more complex than a tax. Perhaps the most common policy distortion involves quantitative restrictions on the supply of capital or controls on interest rates; these controls are usually linked to macroeconomic distortions. The most common market failure involves institutional imperfections, in which capital markets for particular regions or sectors are isolated from one another. Interest rates set by the government and market failures destroy the linkage between observed interest rates and the private rate of return. Instead, rates of return depend on the allocation of financial capital among investors and are likely to reflect excess profits as well as the cost of borrowing. The social rate of return could be higher or lower than the private rate of return.

In this circumstance, other sources of information about the return to capital investment are needed. One source is specific research studies of the return to investment. In national aggregate studies, attempts are made to estimate the value of the economy's total stock of capital equipment. Total national income is then allocated between labor and capital. The income for capital divided by the value of capital stock is a rough measure of the rate of return to capital. Industry-specific studies proceed in a similar manner. The value of the capital stock is compared to an income measure, such as profits plus the annual payments to capital. Comparisons across a number of these industry studies can provide a rough idea of the private rate of return in the economy. Ratios of rental rates to sales prices of agricultural land may give an indication of agricultural sector rates of return, because land usually is a non-depreciating asset. But such comparisons depend on the absence of substantial demands for speculative or consumptive purposes; these demands could cause land prices to exceed the capitalized value of rents. As in the case of exchange-rate adjustments, researcher calculations can be supplemented by expert advice from macroeconomists or institutions involved with investment projects, such as the country's ministry of planning or the World Bank. Box 11.2 provides an illustration of a rate of return calculation.

When information sources are unavailable or unreliable, rough rules of thumb must be used. Countries with higher per capita income levels usually have larger amounts of capital stock. If rates of return to investment behave similarly across countries, higher-income countries will have lower rates of return to investment. Because their capital stocks are small, low-income countries have unrealized investment opportunities that are lucrative compared to those in high-income countries. Empirical estimates of real rates of return are usually in the ranges of 10 to 15 percent for low-income countries, 6 to 10 percent for middle-income countries, and 2 to 6 percent for high-income countries. Of course, the application of such rules of thumb is highly discretionary, and substantial emphasis must be given to the results of sensitivity analyses under such circumstances.

Labor

The greatest complication for labor market evaluations involves recognition of the many types of labor and the choices of private market prices to represent differences in sex, age, and skill levels. Once these choices are made, the price-equivalent value of divergences is added to generate estimates of social labor costs.

The identification of labor market failures usually begins with comparisons of regional wage rates, specified by sex and type of worker, over time. If wage levels for similar types of labor vary substantially among regions or if wages change in a very different manner over time,

Box 11.2. The Rate of Return to Capital in Portuguese Agriculture

The problem is to determine the 1983 rate of return to capital in Portugal. Markets for financial capital are highly distorted in Portugal. The control of credit has been an important tool of macroeconomic policy; by limiting total credit supplies, the government has been able to limit aggregate demand and inflation. At the same time, concern for the borrowers' ability to pay has led to interest-rate controls. These controls have allowed negative real interest rates for borrowers in many of the years preceding 1983. In these circumstances, financial markets are not a useful source of rate of return information.

Instead, use was made of aggregate economic studies of gross domestic product (GDP) distribution and studies that evaluated the stock of capital. The following data were assembled for the 1978–1981 period.

Year	Gross domestic product (billion escudos)				Value of capital stock	Average rate of return
	Total	Labor	Land	Capital		
1978	787.3	450.3	24	313.0	2,230	14.0%
1979	993.3	541.3	30	422.0	2,809	15.0
1980	1,235.0	680.0	35	520.0	3,478	15.0
1981	1,465.4	821.6	42	601.8	4,201	14.3

These data can be used to estimate the marginal rate of return to capital. Under competitive conditions, the ratio of marginal to average value products is equal to the share of the factor in total income. Capital's share of GDP was about 40 percent. Marginal rates of return are therefore 40 percent of the average rate of return; 6 percent is the estimate of the private marginal rate of return.

This rate is then adjusted to account for the impact of various distortions. Because investments were heavily influenced by government directives, alternative investments with potentially higher rates of return were excluded from the capital market. If credit were allocated competitively, rates of return would increase. The effect of this distortion was assumed to equal two percentage points; 8 percent was thus the estimated social marginal rate of return.

The next adjustment to the private rate of return involved accounting for capital market segmentation. In Portugal, low interest rates on savings discourage the use of savings accounts; people are encouraged to invest directly in businesses or assets. Because small-farm agricultural families are major recipients of emigrant remittances and because large farms enjoyed substantial profits, 4 percent was chosen as the private marginal rate of return in agriculture.

Finally, the private rate of return had to be adjusted for producers who had access to special lines of credit. These lines of credit subsidized the costs of borrowing between two and six percentage points. The unsubsidized borrowing rate in 1983 was 28.5 percent. But if the loan was used for particular types of machinery or other farm equipment, borrowing rates were between 22.5 and

26.5 percent, depending on the input. Because such loans required substantial numbers of forms and long time periods for evaluations, the transactions costs for such loans were high for borrowers. Two percentage points were used as the average net value of the subsidy.

In summary, three private rates of return were used in the analysis: 2 percent for users of subsidized agricultural credit, 4 percent for unsubsidized agricultural producers, and 6 percent for producers outside the agricultural sector; 8 percent was used as the social rate of return. These numbers are approximations at best. Although the analyst might have some confidence in the identification of divergences and the direction of their effects on the rate of return, exact quantification is implausible in most cases. Because of the approximate nature of the adjustments, sensitivity analysis of the PAM results, based on a range of assumptions for the private and social costs of capital, was an important portion of the subsequent analysis.

---

nonessential market fragmentation and monopsony power may be present. But wage-rate comparisons are not by themselves sufficient to justify adjustment of private prices. First, the cost of living can differ across regions, and real wage differences can be much less than nominal wage differences. Second, wages may adjust only slowly across regions, because labor does not respond instantaneously to changes in relative earning opportunities. Third, migration from region to region is costly. If relative wages rise in a particular region, labor from another region may choose not to migrate to the new higher-wage area because the costs of migration are larger than the net gains. Therefore, even if labor markets are well integrated, regional wages can be expected to demonstrate some independence.

If any of these circumstances can account for wage differentials, no adjustment of private wages is necessary to approximate social prices for labor. But if market imperfections are present in a particular region, social wage rates will be somewhere between those in the monopsony power region and those in the other regions of the country. When monopsonistic regions use a small share of the total labor market, this equilibrium wage will lie very close to the wage in the non-monopsony region, which can then be used as an estimate of the social price without significant error.

Policy distortions usually entail legislated wage rates or taxes on and subsidies on the use of particular categories of labor. Evaluation of distortions must determine whether the regulations are binding. When the labor market completely ignores a legislated minimum-wage rate, for example, the private wage equals the social wage; both prices are

Box 11.3. The Social Wage Rate in Portuguese Agriculture

Investigations about the social wage rate begin with inspection of agricultural wages by sex and region. The following table lists daily agricultural wage-rate data for Portugal for the 1978–1982 period (in escudos).

	National average		Regional average for males		
Year	Females	Males	South	North	Central
1978	161	227	193	239	240
1979	191	278	242	295	287
1980	232	324	299	331	343
1981	331	438	460	478	451
1982	420	545	496	582	535

The data appear to be consistent for an integrated labor market. Wages for females and males nationally and for within regions appeared to move in a similar manner. However, relative wages are by no means constant. North and central region wages began at similar levels; by the end of the period, northern wages were 9 percent higher. But these differences are not large relative to costs of migration. Further, no evidence of migratory barriers could be found; to the contrary, indications of mobility were common. For example, hired female labor was found to substitute to an increasing extent for tasks that were traditionally the domain of hired male labor. These adjustments reflected the growing scarcity of hired male labor.

The chief distortion in the labor market involved the effect of social security payments, mandated by law after 1974. Because most agricultural employers did not pay such taxes, private and social wages were considered equal for agriculture. The key issue was determination of the effects of legislation on the industrial and service sector wages.

below the minimum wage. If only some sectors observe the regulations, the wages in the unregulated sector provide a measure of the social price. Private prices used in the budget will vary according to whether producers in the commodity system observe the minimum wage. Within a commodity system, some activities could observe the regulation while others do not, so private wages can differ among activities.

Treatment of employer-paid taxes on labor follows a similar procedure. The issue is whether the regulation has actually raised the reward to labor or employers have simply lowered the money wage so that workers' total compensation remains unchanged. The presence of a globally binding regulation should be associated with some unemployment. If legislated wages are associated with full employment and all employers pay the tax, the analyst can assume that the legislation is

The following table compares indexes of real wages for agriculture and manufacturing.

	Real wage index (1975 = 100)	
Year	Agriculture	Manufacturing
1974	98	80
1975	100	100
1976	97	104
1977	88	97
1978	81	90
1979	80	87
1980	80	94
1981	90	94
1982	92	93
1983	86	88

Although the index values do not change identically in each year, wage levels over time move in roughly similar patterns. The principal exception to this generalization involves the jump in real manufacturing wages during 1974–1975. This change was not matched by the agricultural sector, and it coincided with the institution of the social security laws. Moreover, urban unemployment became an increasingly serious problem during the period, suggesting that private market wages were above their social level. For these reasons, social labor costs in the industrial sector were evaluated net of social security taxes; private labor costs included these taxes, a difference of about 24 percent. These adjustments applied almost exclusively to postfarm activities.

nonbinding. The private cost of labor then equals its social cost (money wages plus employer-paid taxes on labor use). Alternatively, full employment coinciding with large sectors of the economy that do not pay the legislated costs suggests that private and social costs diverge. Less labor is employed in the regulated sector; the unemployed portion is diverted to the unregulated sector, forcing down wages. Social wage rates then lie between observed wages in the regulated and unregulated sectors. Box 11.3 illustrates the procedures for determination of the social wage for agricultural labor in Portugal.

Land

Land is unique because it is the only truly fixed factor in agriculture. In suburban locations, agriculture might not be the only use for land, and prices and rental values will be influenced by off-farm opportunities. But in most areas, the only alternative to agricultural use is no use at all (if forestry is included as an agricultural activity). In these cases, land acts as a residual claimant on the profits from farming.

Divergences that affect the prices of agricultural outputs and nonland inputs have a direct impact on the rental value of land. If the prices of the principal outputs of a region increase, profits will increase. Ultimately, land values will increase because individual producers are willing (and able) to pay an increased amount for the right of access to farmland. Indeed, if agriculture is a price-taker in all other input markets (because agricultural demands for labor and capital are a small share of the total economy's demand for these factors), the rental price of land will absorb all of the change in the profitability of the farm activity. Only if arable land supplies are in surplus will the price of land remain unaffected (and presumably near zero).

To draw conclusions about the effects of policy distortions and market failures on the choice of agricultural activities, the social land rental value is usually measured as the value of the land in its most profitable alternative use. If oat production represents the only alternative to wheat production, for example, the social cost of land for the wheat activity is represented by the social profits (excluding land) from the oat activity. If the wheat activity did not generate returns at least as high as those available to oat production, farmers would choose to use their land to produce oats.

When private and social land are included in measures of domestic factor costs, the measure of factor market costs of divergences ($K = C - G$) requires careful interpretation. Because both the social and the private value of land are determined in relation to alternative uses, K will include some effect of the policies and market imperfections that influence the profitability of alternative crops. Moreover, the alternative crop might not be the same in the social and private cases. A second set of influences involves direct distortions in the land market. Governments sometimes try to alter the distribution of profits between tenant and landowner by imposing controls that limit the land rents paid by tenants. If such controls are binding, private market rental rates will be less than the full amount of private profit.

Two caveats may modify the use of social profits in alternative activities as proxies for the social rental rate. First, profits may include returns to some inputs not evaluated in the budget, such as managerial skill. One farm activity may show higher profits than another activity, but the difference may be explained by managerial skill. The returns to land then would vary little between the two systems. Second, systems can differ substantially in terms of riskiness, and variation in profitability may be important to the activity choice. For example, vegetable crops often provide higher returns on average than staple food crops. Yet many producers continue to grow food crops because of the greater stability of returns from year to year. In this circumstance, land values will not rise so high that staple crop production will be eliminated, and land of identical quality will produce a variety of crops.

If risk effects and managerial requirements are thought to differ substantially among alternative systems, one must investigate the system alternatives as an explicit component of the analysis. By considering all major commodity systems in a particular region, the analyst can directly compare the relative profitabilities of alternative land uses without incorporating the rental cost for land into the PAM. The construction of the PAM proceeds without consideration of land costs, and domestic factor costs (C and G) include only labor and capital costs. Systems that do not generate revenues sufficient to pay the costs of tradable inputs, labor, and capital will be unattractive in the long run. But so may systems that offer low profits relative to alternative activities. Box 11.4 illustrates the two approaches to profitability calculations with examples from irrigated land in northwest Mexico and rainfed land in Portugal.

Estimating Input Use of the Commodity System

The last set of adjustments to private costs and returns involves accounting for the possible response of producers to the social prices of outputs and inputs. Patterns of input use by a profit-maximizing firm are dictated by consideration of marginal cost. The producer compares this value to marginal revenue when increases or decreases in output are contemplated. The general case, in which marginal returns to input use are diminishing, is illustrated in Figure 11.3. Figure 11.3a shows the relationship between the use of fertilizer input and wheat output, given by curve ABCD. Fertilizer is only one of the inputs used in production, and the entire production process can be represented by a family of similar curves, one for each input. Producers using fertilizer will be concerned with the relationship between the marginal change in the total cost of inputs and the incremental gain in output that results from increased input use. The producer will insist that

$$(\Delta Q_W)(P_W) \geq (\Delta Q_F)(P_F)$$

or

$$\frac{\Delta Q_W}{\Delta Q_F} \geq \frac{P_F}{P_W}$$

Box 11.4. The Social Value of Land

Wheat in Portugal

The only alternatives to dryland wheat production in the Alentejo region of Portugal are oats and barley. Both crops require managerial and cultivation practices similar to those of wheat. The crops do not appear to differ much in terms of profitability risk. The private land rental rate is 600 escudos per hectare. Rent control laws require a lower rental rate, but these laws are not enforced.

In calculations of the social value for land, barley represents the best alternative crop. The social costs and returns to 1 hectare of barley are summarized in the following table.

Revenues	
1,500 kg barley @11 esc/kg	16,500
100 bales straw @50 esc/bale	5,000
	21,500
Costs	
Labor (© 80% of private wage)	1,700
Capital (@ 30% above private cost)	5,445
Fertilizer	5,272
Other tradables	4,111
	16,528
Social profit	4,972

The social profit of barley production is estimated as 4,972 escudos per hectare. This price is thus used as the social value of land.

The left-hand side of this inequality is the slope of the input-output productivity curve, ABCD, and the right-hand side is the input-output price ratio. In Figure 11.3a, this relation means that the point of tangency between the price line and the productivity curve (point B) represents the point of maximum profit. Because marginal productivity declines as input use increases, expansion of input use beyond that associated with point B will decrease net profits; the additional cost of fertilizer ($\Delta Q_F \times P_F$) is greater than the additional value of wheat produced ($\Delta Q_W \times P_W$). Only if the relative price of wheat increases will fertilizer use increase. In Figure 11.3a, an increase in the wheat price from P_w^1 to P_w^2 causes the price line to become less steeply sloped. Optimum input use increases, indicated in the graph by point C.

An increase in the wheat price will exert similar effects on the use of other inputs in wheat production; the aggregate impact of all these

Irrigated Land in Sinaloa, Mexico

The social valuation of irrigated land in northwest Mexico provides a sharp contrast to the valuation of dryland areas of Portugal. In northwest Mexico, a wide range of crops is technically feasible, with differences in market destination (export or domestic), managerial requirements, and price variability. In this case, the estimation procedures compare profits before land cost for as many crops as possible (see the following table). Private market land rental rates were 40,000 pesos per hectare for all crops except vegetables, which were 60,000 pesos per hectare.

Crop	Social profits before land cost (pesos per hectare)
Corn	52,000
Wheat	22,600
Rice	66,200
Beans	−6,900
Sorghum	40,900
Soybeans	−800
Safflower	−13,800
Large tomatoes (export)	2,206,000
Cherry tomatoes (export)	1,232,000
Green bell peppers	1,150,000
Cotton (export)	−16,400

These results show the enormous increase in net returns for exporters of vegetable crops, but these crops require conditions of financial and production management that are very different from the conditions for other field crops. Because of dissimilar management requirements and riskiness, differences in the costs of land would not be expected to account for all of the extra profits in vegetable and field crop production. For farmers without access to vegetable crop production, four of the crops—beans, soybeans, safflower, and cotton—offer negative returns. Social profits in the remaining crops—corn, wheat, rice and sorghum—range from 22,000 to 66,000 pesos per hectare. Social land values would probably fall somewhere in this range.

changes on output is summarized in the marginal cost curve of the firm (Figure 11.1b). Because all input productivities are assumed to diminish with increasing amounts of input, the marginal cost curve takes on an increasingly steep slope as price increases and output expands. The proportional impact on output of output price increases becomes smaller, because increases in input use have reduced incremental effects on production.

Both the average cost curve and the marginal cost curve can be used to measure profits. In terms of average cost, total profits in Figure 11.3b

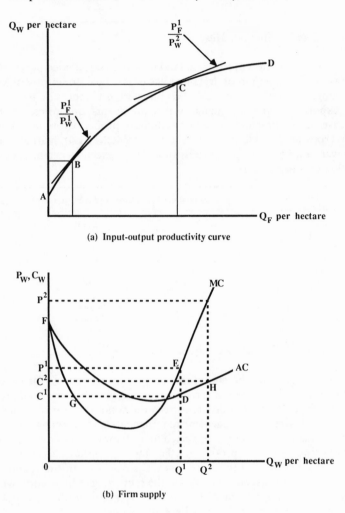

(a) Input-output productivity curve

(b) Firm supply

Figure 11.3. Profit maximization with diminishing marginal productivity of inputs

are the rectangle P^1C^1DE; in terms of marginal cost, profits are the area of rectangle P^1EQ^10 less the area under the marginal cost curve, FGE. Because the average cost curve does not determine producer response to changes in output or input prices, old budgets become inaccurate portraits of producer profitability and input use as prices change. If P^1 represents the private market price for wheat and P^2 represents the social price, Figure 11.3b shows that an increase in wheat price from P^1 to P^2 induces an increase in wheat output from Q^1 to Q^2. This output increase in attained by increased use of at least one input; because the

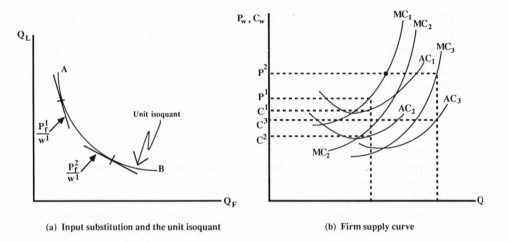

(a) Input substitution and the unit isoquant   (b) Firm supply curve

Figure 11.4. Input substitution and output supply

productivity of input use is diminishing, average cost per unit of output increases. Social costs per unit of output are thus larger than private costs. The correct social cost corresponding to social output price P^2 would be that associated with point H; social profits would be $P^2 - C^2$ per unit of output. If budgets associated with point D were used to calculate social profitability, the estimate of profit per unit of output would be $P^2 - C^1$, overstating the true measure of social profit.

To this point, input prices have been assumed to remain constant as output prices change. But changes in input prices can also induce changes in input use. The profit-maximizing producer tries to decrease usage of inputs whose prices increase in relative terms and increase usage of inputs whose prices fall. These changes create a new pattern of input-output relations in a manner analogous to the previously described effects of output price changes. In this case, however, changes in input prices cause shifts in the firm's marginal and average cost curves rather than movements along these curves. Each new marginal and average cost curve entails input-output relationships for social cost and return calculations that differ from those used in the private cost calculations.

Figure 11.4 illustrates the impact of input substitution on the firm's cost curves. Figure 11.4a shows the unit production isoquant. The curve AB represents all combinations of inputs (in this case, labor and fertilizer) that can be used to produce one unit of output (wheat). Under constant returns to scale, this unit isoquant is sufficient to describe all possible input combinations; isoquants representing higher levels of output will be direct replications of the unit isoquant. To minimize production costs, the producer seeks the point of tangency between the unit isoquant and the line having a slope equal to the factor price ratio (by reasoning similar to that used for the input-output productivity curve).

The diagram illustrates the impact of a decrease in fertilizer price (because the private market price of fertilizer exceeds the social price). This price change causes a shift in the optimal input combination to relatively more fertilizer and less labor. The input price change encourages an expansion in firm output as well; these changes are reflected in the input-output productivity curve of Figure 11.3. But now output expansion occurs as a result of declines in input costs rather than an increase in output price; therefore, cost curves must shift rather than remain fixed.

These shifts are described in Figure 11.4b. The cost curves MC_1 and AC_1 correspond to the initial input prices of P_F^1 and w_L^1. Private revenues per unit are P^1, private costs are C^1, and private profits are $P^1 - C^1$. An initial effect of reducing the fertilizer price from P_F^1 to P_F^2 is to reduce the cost of every fertilizer-using input combination that produces wheat. This effect is shown in the diagram as direct downward shifts of the marginal and average cost curves, to MC_2 and AC_2. Estimated social cost, C^2, is less than private cost. But two further changes occur if private prices are altered to social prices: new input combinations are used because of the relatively lower price of fertilizer, further shifting the cost curves to MC_3 and AC_3; and increased amounts of all inputs per unit output are used because of the change in output price.

As in the case described in Figure 11.3, observed input-output relationships provide a misleading estimate of social profits. If the analyst first uses budgets that reflect private market incentives and then changes input prices from private to social values, social costs per unit will be estimated as C^2 and social profit per unit will be $P^2 - C^2$. But true social costs, measured after all the incentive effects of social prices are incorporated, are C^3, and social profit is $P^2 - C^3$. In this illustration, true social profit per unit of output is less than that estimated from the initial budgets. However, a larger shift from MC_2 to MC_3 could create the opposite situation, in which true social profits exceed those estimated from budgets. No predictable direction of bias prevails between true social profit and estimates based on observed input-output relationships.

Empirical estimation of the input and output responses can be made with econometric models of supply response and input demand for commodity systems; Box 11.5 illustrates the approach. Changes from private to social prices of outputs and inputs yield estimates of the quantities of inputs and outputs that coincide with social price incentives. Social revenues, costs, and profits (the second line of PAM) are then calculated by multiplication of social quantities and social prices. The impacts on the firm of distortions and market failures (the third line of the PAM) then contain both quantity and price effects. However, the construction of econometric models can be very demanding of data and research resources. The estimates from econometric models are also subject to uncertainty, and ignoring quantity effects altogether can be preferable to using the results from a poor econometric model.

The decision to undertake econometric estimation hinges on the capability of fixed coefficient assumptions to yield results that are close to the true measures of social revenues and costs. When input-output relationships are fixed, the use of observed input-output relationships provides exact measures of social revenues, costs, and returns. This circumstance is illustrated in Figure 11.5. The input-output productivity curve is represented by a single linear segment because marginal productivities are a constant (Figure 11.5a). Alternative combinations of inputs are not feasible (Figure 11.5b), and the average and marginal cost curve is a straight line (Figure 11.5c). In these circumstances, changes in output prices (increases in the price of wheat) and changes in input prices (reductions in the price of fertilizer) have no impact on the optimal tangency points. Social profit is measured exactly as $P_w^2 - C^2$. This case provides obvious advantages to the policy analyst. One set of observable input-output coefficients fully describes the technology side of the analytical problem, and attention can be focused on the collection of price-related information. The decision to use such an approximation of reality depends, of course, on conditions particular to the commodity of interest.

If necessary, an analytical compromise can be struck between the extreme alternatives of continuously diminishing marginal returns and fixed input-output relationships. If the input-output productivity curve is given in linear segments, marginal input productivities change in discrete steps. Marginal productivities are assumed to be constant over some range of input use but can change from one interval to the next. Figure 11.6 illustrates this procedure for one input (fertilizer) and one output (wheat). In Figure 11.6a the relationship between fertilizer input and wheat output is described by the line ABCDE; this line is made up of four linear segments. At low levels of input use, such as an amount

Box 11.5. The Calculation of Social Input-Output Coefficients

The estimation of likely producer response to changes in output and input prices requires a comprehensive estimation technique, such as the profit function. In most cases, cost and supply functions will not provide sufficient information to estimate all of the relevant interactions between inputs and outputs. The application of profit functions to the estimation of social input-output coefficients is illustrated here for a simple one-output (rice), two-input (labor and fertilizer) model.

Private prices are assumed to be 200 Rp per kg of rice, 200 Rp per kg of fertilizer, and 1500 Rp per day of labor. These prices are associated with yields of 6,000 kgs rice per hectare, fertilizer input use of 450 kgs per hectare, and labor input of 300 days per hectare.

Suppose that social values are determined as 150 Rp per kg of rice, 220 Rp per kg of fertilizer and 1,500 Rp per day of labor. Estimation of social input-output coefficients must consider two categories of responses. First, the decrease in output price encourages a decline in yield, that corresponds with a reduction in demand for fertilizer and labor inputs. These changes are shown as a movement from point A to point B in the inset Figure. Second, the increase in fertilizer price encourages a decrease in fertilizer use and an increase in the use of substitute inputs (labor). Together, these changes in input combinations may imply an additional change in yield. These changes are shown as a movement from point B to point C. In constructing the PAM, private cost and returns are associated with point A, whereas the social values are associated with point C.

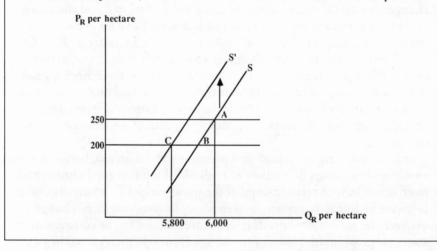

All necessary information for estimation of input and output quantities associated with point C can be provided by a profit function. Use of a trans-log profit function is assumed to generate the following elasticity values: output with respect to output price, 0.13; fertilizer demand with respect to output price, 0.12; labor demand with respect to output price, 0.34; fertilizer demand with respect to fertilizer price, − 0.62; labor demand with respect to wage, − 0.44; labor demand with respect to fertilizer price, 0.10; output with respect to fertilizer input, 0.075; output with respect to labor use, 0.375; the first three elasticities are used to estimate the impact of the 25 percent reduction in output price: estimated rice yield is 5,805 kgs per ha (− .25 x .13 x 6,000 + 6,000); estimated fertilizer use is 436 kgs per ha (− .25 x .34 x 300 + 300).

The next three elasticities are used to calculate the impacts of a 10 percent increase in fertilizer price: fertilizer use declines to 409 kgs per ha (.10 x (−.62) x 436 + 436); labor use increases to 277 days (.10 x .10 x 274 + 274). These are the input quantities associated with the social prices. To approximate the impact of fertilizer price response on yield, the last two elasticities from the above list are used: the decrease in fertilizer causes yield to decline by 27 kgs (− .062 x .0748 x 5805); increased labor inputs cause yield to grow by 22 kgs (.01 x .375 x 5805). On balance, yield falls by 5 kgs, to 5,800 kgs per hectare. The PAM becomes the following:

	Revenue	Tradable Input	Domestic Factors	Profit
Private	1,200,000	90,000	450,000	660,000
Social	870,000	89,980	415,500	364,520
Distortions and divergences	330,000	20	34,500	295,480

If the analyst had ignored the producer response to altered prices and estimated social revenues and costs on the basis of observed (private) input-output coefficients, the following PAM would result:

	Revenue	Tradable Input	Domestic Factors	Profit
Private	1,200,000	90,000	450,000	660,000
Social	900,000	99,000	450,000	351,000
Distortions and divergences	300,000	−9,000	0	309,000

In this example, total transfers (L) are over-estimated by only 4.6 percent, reflecting the prominence of inelastic values for the assumed elasticities; the errors in individual categories (I, J, and K) are much larger.

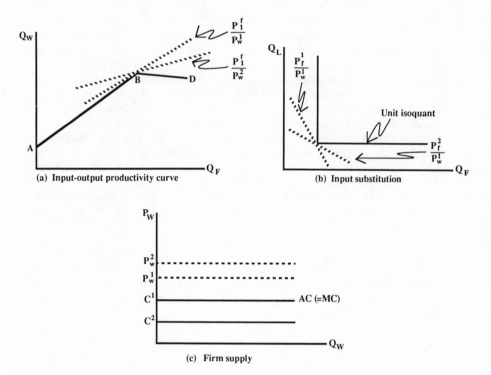

Figure 11.5. Firm behavior when input-output coefficients are fixed

between 0 and Q_F^1, marginal response of output to fertilizer input is larger than at higher levels of input use, such as Q_F^3. Segment AB thus has a steeper slope than segments CD and DE. Marginal productivities are constant within each interval but diminish steadily across intervals that correspond to higher levels of input use.

Facing five linear segments, the producer chooses among only five alternative levels of fertilizer use: 0, Q_F^1, Q_F^2, Q_F^3, and Q_F^4. Points in between will never be chosen. The explanation for this choice pattern relates directly to profitability considerations, represented by

$$\frac{\Delta Q_W}{\Delta Q_F} \geq \frac{P_F}{P_W}$$

For most price ratios, the tangency of the price line and the productivity curve corresponds to one of the kinks in the productivity curve. For example, all prices between

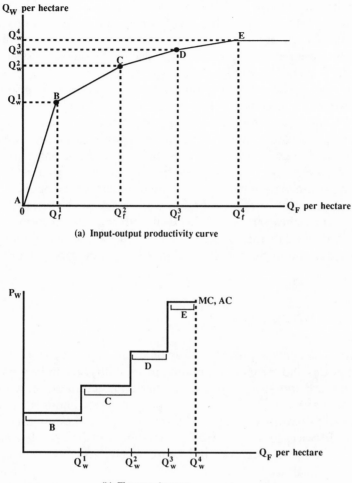

(a) Input-output productivity curve

(b) Firm supply curve

Figure 11.6. Input use, technology choice, and output under piecewise linear productivity curve

$$(Q_W^2 - Q_W^1) / (Q_F^2 - Q_F^1) \text{ and } (Q_W^3 - Q_W^2) / (Q_F^3 - Q_F^2)$$

dictate the selection of point C as the maximum profit level of fertilizer use. This result follows directly from the linearity of the productivity curve. Within each interval, subsequent increases in input use contribute equally to profitability.

Figure 11.6b illustrates the firm supply curve under the piecewise linear productivity curve. As the price of wheat increases (with the price of fertilizer held constant), the producer responds by using larger quantities of fertilizer per hectare. Output per hectare increases in discrete steps. Because marginal productivities decline among successive intervals, the steps of the supply curve become progressively larger.

This method of treating diminishing marginal returns means that knowledge of a discrete number of alternative technologies—often obtained from engineering studies, experiment station results, or experiences of other countries—is sufficient to measure social profit. As output prices change from private to social levels, the socially profitable technology may (or may not) differ from the technology currently used (facing private market prices). By evaluating alternative budgets under social price incentives, the profitability of change can be assessed. Analogous approaches can be used to deal with input substitution; linear approximations to the production isoquant can be made, and minimum-cost technologies associated with social input prices can be identified.

Sensitivity Analysis

Sensitivity analysis provides a way of assessing the impact of changed assumptions and errors in estimating profitability. It can be applied to both private and social estimations. In private estimations, it usually involves partial budgeting. In principle, all social parameters can be subjected to sensitivity analysis. However, the social estimates of long-run world prices for output, the cost of labor, and the cost of capital are usually the most uncertain and hence receive the most attention in sensitivity analysis.

The choice of social prices for outputs and inputs is subject to analytical imprecision in several areas. First, estimates of price-equivalent impacts of factor market divergences might not be much better than educated guesses, especially for rates of return to capital and short-run effects of distorted foreign exchange rates. Second, divergencs additional to factor market divergences may influence domestic factor prices, and their impacts may not be well-understood. For example, widespread protection to outputs that are intensive in a particular factor will probably elevate that factor's price. Third, price response within the commodity system could cause the quantities of inputs employed under social prices to be different from those used in the estimation of private profits.

One approach to sensitivity analysis involves the calculation of breakeven values for social profitability. The breakeven value of a parameter is the value necessary to achieve zero social profit when all other revenues and costs are held at their initial values. A second indicator is the elasticity of social profitability with respect to a particular parameter; it is expressed as the ratio of the percentage change in social profit of the system relative to the percentage change in the parameter. The calculation of these elasticities proceeds by an increase in the parameter of interest by an arbitrary percentage (for example, 10 percent). Social profitability is then recalculated and compared to the initial value to estimate the percentage change in social profit. The ratio of the two percentage changes gives the elasticity estimate. Input costs will have negative elasticity values, whereas output prices will have positive elasticity values. The larger the value of the elasticity, the more sensitive are the results to measurement error or parameter change in the social evaluation exercise.

Interpretation of the results of sensitivity analysis is somewhat arbitrary. Whether elasticity values are large or breakeven values are very different from initial values depends on the quality of the initial estimations and the degree of potential change in the variables. As a rule of thumb, if breakeven values differ by less than 15 percent from their initial values, the analyst should be cautious about associating positive or negative values of social profitability with the commodity system. In these instances, judgments about the desirability of the system in the economy may have to be based more on nonefficiency objectives, such as income distribution, food security, and regional development impacts. The tradeoff between efficiency and nonefficiency goods remains, but more empirical research is needed before quantitative estimates are useful. If, however, results appear robust following sensitivity analysis, efficiency gains or losses should be a significant element in policy decisions concerning the commodity system.

Concluding Comments

All social price calculations rely to some degree on the judgment of the analyst. Principles for the determination of appropriate world and domestic factor prices are relatively easy to establish, but their implementation inevitably is limited by data availability. Some would argue that this problem provides sufficient grounds to avoid social price calculations altogether and to focus economic analyses instead only on issues that can be directly addressed by available data. Alternatively, the logic underlying social price calculations could be altered by the use of a new definition of optimality that associates optimal conditions with data that are easier to collect. An example is a type of second-best approach that assumes that all divergences external to the commodity system are beyond the influence of policy-makers. Divergences in factor and tradable-input markets are ignored, and social input prices are assumed to be equal to private prices.

But such approaches are not especially helpful in most policy analyses. Economists do not determine the issues of economic policy; policy-makers and societal interest groups do, and policy-oriented empirical analysis are expected to address these issues as comprehensively as possible. Trying to hide difficult to measure parameters under the cloak of arbitrary definitions of optimality does little to clarify the economic impacts of governments on agricultural producers. Like economic theorists, empirical analysts desire to minimize the number of assumptions needed to generate results. But information is never perfect, and assumptions form an inevitable element of applied analysis. If analysts provide full descriptions of the procedures and assumptions they have used, subsequent researchers will have ample opportunity to improve upon results.

Bibliographical Note to Chapter 11

Empirical estimations of shadow prices require assumptions and approaches that are different for each investigation, because available data vary widely across countries and commodity systems. Often, the details of such calculations are omitted or treated only summarily when materials reach the publication stage. An early study that gives the flavor of such exercises is M. F. G. Scott, J. D. MacArthur, and D. M. G. Newbery, *Project Appraisal in Practice* (London: Heinemann, 1976); this work applies the Little-Mirrlees method to analyses of projects in Kenya. Works that describe social pricing exercises in a domestic resource cost methodology include Scott R. Pearson et al., *Rice in West Africa: Policy and Economics* (Stanford, Calif.: Stanford University Press, 1981); and Walter P. Falcon et al., eds., *The Cassava Economy of Java* (Stanford, Calif.: Stanford University Press, 1984). The exposition that is closest to the methodology discussed here is Scott R. Pearson et al., *Portuguese Agriculture in Transition* (Ithaca: Cornell University Press, 1987).

Social pricing of tradable commodities almost always begins with world prices. Prices for many agricultural commodities are monitored by several international organizations. The World Bank's *Commodity Trade and Price Trends* (Washington: World Bank) contains prices for commodities and for several inputs. Other sources of world prices are the International Monetary Fund's *International Financial Statistics* (Washington: International Monetary Fund), the UN Food and Agricultural Organization's *Monthly Bulletin of Food and Agricultural Statistics* (Rome: Food and Agricultural Organization), and various publications of the U.S. Department of Agriculture, the U.K. Commonwealth Secretariat, and several international commodity organizations (for coffee, cocoa, sugar, cotton, rubber, wheat, olive oil, and others).

World prices are adjusted to correspond to the particular market characteristics for the commodity system under study. A review and analysis of the role of quality and hedonic pricing techniques is provided in Angus Deaton and John Muellbauer, *Economics and Consumer Behavior* (New York: Cambridge University Press, 1980), chap. 10. An approach to the evaluation of quality effects on price is provided in Eric Monke and Todd Petzel, "Market Integration: An Application to International Trade in Cotton," *American Journal of Agricultural Economics* 66 (November 1984): 481–87.

The cif-fob distinction is most relevant for processed commodities or for countries with large transportation costs. Analyses of the importance of this distinction are provided in Eric Monke, S. R. Pearson, and J. P. Silva-Carvalho, "Welfare Effects of a Processing Cartel: Flour Milling in Portugal," *Economic Development and Cultural Change* 35 (January 1987): 393–407; Eric Monke, "The Economics of Rice in Liberia," in Pearson et al., *Rice in West Africa*, pp. 141–72; and John McIntyre, "Rice Production in Mali," in Pearson et al., *Rice in West Africa*, pp. 331–60. The problem of price variability over time may be approached as an insurance problem. This literature is reviewed in Peter Hazell, C. Pomareda, and A. Valdes, *Crop Insurance for Agricultural Development* (Baltimore: Johns Hopkins University Press, 1986). Also relevant is the futures market literature on options contracts; see Todd Petzel, "Alternatives for Managing Agricultural Price Risk: Futures, Options and Government Programs," American Enterprise Institute *Occasional Paper* (November 1984). An example of the hazards of forecasting expected prices are apparent from ex post analyses of almost all projection exercises; compare, for example, the rice prices reported in Box 13.1 with those expected by Walter P. Falcon and Eric Monke, "International Rice Trade," *Food Research Institute Studies* 17 (1979–1980), 279–306.

Chapter 7 of W. M. Corden, *Trade Policy and Economic Welfare* (Oxford: Clarendon Press, 1974), summarizes the arguments about optimal tariffs. Much of the empirical work on optimal trade taxes in agriculture has been done by Andrew Schmitz and colleagues. Examples of this work are Colin Carter and Andrew Schmitz, "Import Tariffs and Price Formation in the World Wheat Market," *American Journal of Agricultural Economics* 61 (August 1979): 517–22, and Andrew Schmitz et al., *Grain Export Cartels* (Cambridge, Mass.: Ballinger, 1981). Another work that discusses (skeptically) empirical possibilities for price controls is Carl van Duyne, "Commodity Cartels and the Theory of Derived Demand," *Kyklos* 28, no. 3 (1975): 597–612.

Discussions of social prices for domestic factors are contained in the books cited in the first paragraph. In the factor markets, returns to capital are usually the most difficult to calculate, even in developed countries. An example of the empirical complexities of rate of return calculations is provided in Robert M. Coen, "Alternative Measures of Capital and Its Rate of Return in United States Manufacturing," in *The Measurement of Capital*, ed. Dan Usher (Chicago: University of Chicago Press, 1980), pp. 121–49. An application to a developing country economy is contained in Sergio Bitar and Hugo Trivelli, "The Cost of Capital in the Chilean Economy," in *Analysis of Development Problems*, ed. R. S. Eckaus and P. N. Rosenstein-Rodan (Amsterdam: North-Holland, 1973), pp. 147–66. Most of the research on the relationship between land prices and land rents has been performed with U.S. data. Studies find that farm-based returns vary in importance along regions as determinants of prices. In some cases, expectations and intrinsic value cause apparent rates of return to land purchases to fall below the rates of return to other capital investments. Recent studies include Tim Phipps, "Land Prices and Farm-Based Returns," *American Journal of Agricultural Economics* 66 (November 1984): 422–29; C. Arden Pope, III, "Agricultural Productive and Consumptive Use Components of Rural Land Values in Texas," *American Journal of Agricultural Economics* 67 (February 1985): 81–86; and Lindon J. Robison, D. A. Lins, and R. Ven Kataraman, "Cash Rents and Land Values in U.S. Agriculture," *American Journal of Agricultural Economics* 67 (November 1985): 794–805.

Estimations of the factor price impacts of output price distortions and input substitution have relied largely on assumptions. A review of some of the empirical work is provided in Bela Balassa, "The Interaction of Factor and Product Market Distortions in Developing Countries," *World Development* 16 (1988): 449–63. Computable general equilibrium models have shown some potential in this area. An example is John Whalley, *Trade Liberalization among Major World Trading Areas* (Cambridge: MIT Press, 1985), which analyses the factor price impacts of removing trade barriers. An example of a simplified single country model is provided in Kym Anderson and Peter G. Warr, "General Equilibrium Effects of Agricultural Price Distortions: A Simple Model for Korea," *Food Research Institute Studies* 20 (1987), 245–63.

Recent work on producer behavior has been developed in terms of profit functions. A great attraction of this approach is its use of output price, output quantity, and input price data to estimate simultaneously both input demand and output supply response. General discussions of this approach are contained in W. E. Diewert, "Duality Approaches to Microeconomic Theory," in *Handbook of Mathematical Economics*, vol. 2, ed. Kenneth Arrow and M. Intriligator (Amsterdam: North-Holland, 1982), and L. J. Lau, "Applications of Profit Functions," in *Production Economics: A Dual Approach to Theory and Applications*, ed. M. Fuss and D. McFadden (Amsterdam: North-Holland, 1978), pp. 133–216. Some useful empirical applications of profit functions are S. S. Sidhu and C. A. Baanante, "Estimating Farm-Level Input Demand and Wheat Supply in the Indian Punjab Using a Translog Profit Function," *American Journal of Agricultural Economics* 63 (May 1981), 237–46; C. Shumway,

"Supply, Demand and Technology in a Multi-Product Industry: Texas Field Crops," *American Journal of Agricultural Economics*, 65 (1983), 748–60; and J. M. Antle, "The Structure of U.S. Agricultural Technology, 1910–78," *American Journal of Agricultural Economics* 66 (November 1984), 414–21.

The literature on the application of econometric techniques to production behavior is vast. Three texts that provide a general introduction to estimation techniques and procedures are Jan Kmenta, *Elements of Econometrics*, 2d ed. (New York: Macmillan, 1986); Peter Kennedy, *A Guide to Econometrics*, 2d ed. (Cambridge: MIT Press, 1985); and Robert S. Pindyck and Daniel L. Rubinfeld, *Econometric Models and Economic Forecasts*, 2d ed. (New York: McGraw-Hill, 1981). Empirical estimation is almost universal in affirming the output response of developing country producers to changes in both output and input prices. Much of the empirical literature is summarized and referenced in Hussein Askari and John T. Cummings, *Agriculture Supply Response: A Survey of the Econometric Evidence* (New York: Praeger, 1976).

CHAPTER 12

Interpretation and
Communication of PAM Results

THE PRINCIPAL TASK of this chapter is to show how to interpret the
results of the PAM method. Because measures of divergences can include the effects of efficient and distorting policies and of market failures, it is useful to know how much of the difference between private
and social valuations is attributable to each influence. These issues are
examined in the first section of the chapter. The second section illustrates the use of PAM results for agricultural planning analyses of
commodity price, public investment, and agricultural research policies.
But it is desirable to go beyond pure analytics in policy analysis. Effective analysts will draw insights from the results to explain their
meaning and limitations to policy-makers. The final section discusses
strategies for the communication of PAM results to policy-makers.

Interpretation of the Effects of Divergences

Divergences include two types of influences that cause the economy to
use its scarce resources inefficiently so that it does not create the highest
possible levels of income. One type is caused by government policies
that distort the pattern of production, moving it away from the most
efficient use of domestic resources and international trading opportunities. Governments usually enact distorting policies to favor particular interest groups or because they are consciously trading off the
consequent efficiency losses against their perception of such nonefficiency gains as changes in income distribution and improvement in the
countries' ability to feed themselves. The second type of influence arises

226

because certain markets fail to bring about an efficient allocation of goods or services. Market failures are usually far more prominent in factor markets than in product markets.

Output Transfers

An output transfer, I, is defined as the difference between the actual market price of a commodity produced by an agricultural system, A, and the efficiency valuation for that commodity, E. If the system has more than one output, the matrix entries A, E, and I will be made up of the sum of market prices, efficiency prices, and output transfers for all outputs. However, since the actual analysis is constructed on a commodity-by-commodity basis, this discussion assumes that only one output is produced. In most countries agricultural outputs enter into international trade in the absence of trade-distorting policies. For these tradables the appropriate efficiency valuation is given by the world price (fob export or cif import).

The lack of participation in international trade does not in itself mean that the output is nontradable; when a government effectively bans imports of a commodity, no trade will be observed. But this absence of trade is the direct result of the distorting policy. Because most agricultural outputs are internationally tradable, this discussion focuses on tradable commodities. In practice, whether a commodity is tradable or nontradable is an important empirical question. Entries into the E box of the matrix, the social valuations, are thus either comparable world prices for tradable outputs or marginal social costs for nontradable outputs.

Divergences, which cause private valuations to depart from their social counterparts, are always the result of either distorting policies or market failures. Governments can, at least in principle, enact efficient policies that correct the inefficiency influences of market failures. This effort is observed only rarely, because failures in output markets are difficult to identify empirically and are thought to be fairly unimportant (on the basis of sketchy evidence), especially in the context of more pressing economic and social concerns. As a practical matter, therefore, in most contexts the measured effects of divergences in output markets are attributed solely to distorting policy.

Governments choose between two principal policy instruments—trade restrictions and taxes or subsidies—if they want private prices to differ from social values set by world prices. If a government wishes the private price to be above the world price for imported goods (as illustrated in Box 12.1), its policy-makers can either restrict international trade or levy a tax on all production, domestic and imported. Alternatively, if the desire is to lower domestic prices of importables relative to cif import prices, the government has only one choice—to subsidize imports with payments from the treasury. The opposite results apply to exportable outputs.

If all agricultural systems under study have identical outputs, the analyst can compare their output transfers simply by contrasting the absolute sizes of the entries in I for all PAMs within or across countries. For example, the output transfer for one wheat system in Portugal can be compared with that of another wheat system in Mexico—if both systems produce only wheat grain and wheat straw. One needs only an appropriate exchange rate to convert both PAM results to a single currency. However, a comparison of the output transfer for a wheat system with that for a corn system requires construction of a ratio to compare the unlike products. This ratio is the nominal protection coefficient on tradable outputs (NPCO), defined as the private price divided by the comparable world price. If there is a single product in the system, the NPCO is given by the ratio of two PAM output entries, A / E. When more than one output is produced, the average NPCO for all products is found by the adding up of all outputs in private prices and then in social prices and by the formation of a ratio of these two sums. This procedure is illustrated in Box 12.1.

Tradable-Input Transfers

The tradable-input transfers, J, are defined as the difference between the total costs of the tradable inputs valued in private prices, B, and the total costs of the same inputs measured in social prices, F. A private output price above its social price means that policy is providing a positive transfer, causing the production system to realize higher private profits or cover greater private costs than it could without the aid of the policy. This positive transfer has a positive sign in the third row of the PAM. Correspondingly, subsidies on tradable inputs cause production to have greater private profitability. The PAM allows aggregation of all of the effects of divergences, combining those influencing outputs, tradable inputs, and factors.

The principles underlying the interpretation of tradable-input transfers are equivalent to those just set out for output transfers. World prices serve as social valuations of all tradable inputs. Nontradable inputs are decomposed into their component tradable-input and pri-

Box 12.1. Output Transfers in a Portuguese Wheat System

| | Revenues (in escudos per kilogram) | | |
	Wheat grain	Wheat straw	Total
Private prices	23.00	4.42	27.42
Social prices	18.37	4.42	22.79
Effects of divergences	4.63	0.00	4.63
(private prices less social prices)			

In this example, the effects of divergences are entirely the result of distorting policy, not of market failures. The actual policy was a quantitative restriction against imports of wheat, which had an effect equivalent to that of an import tariff of 25 percent: $(23.00 / 18.37 - 1.00) \times 100$ percent. No policies affected the price of wheat straw, a nontradable by-product of wheat grain used for animal feed. If the government had chosen to permit an unrestricted supply of wheat imports, the private (actual market) price would have fallen to the social (cif import) price. At that lower price, the country would have imported more wheat, produced less domestically, and consumed more.

This outcome would have been more efficient than the actual one, because too many domestic resources were used to produce a product that could have been imported more cheaply and because local processors (and ultimately consumers) were forced to pay too much for wheat. In effect, the protectionist policy caused the country to give up some of the potential gains from international trade. To evaluate the effectiveness of this policy, one needs to compare the efficiency losses from producing, consuming, and trading inefficiently with whatever gains might have arisen for the government in pursuing nonefficiency objectives, such as income redistribution (favoring wheat farmers over wheat product consumers) and food security (which would be enhanced if domestic variability in wheat quantities and prices were less than variability on the international market for wheat).

The NPCO permits comparison of systems producing unlike outputs. The NPCO on wheat grain only is given by the ratio of the private price of wheat to the social price of wheat, or $23.00 / 18.37 = 1.25$. This result shows that the country's trade-restrictive policy has permitted the private price to be 25 percent higher than without the policy. The private price could be compared with other single-commodity NPCOs. The NPCO for the entire wheat system is found by formation of a ratio of total revenues in private and social prices. This result, $27.42 / 22.79 = 1.20$, indicates somewhat lesser protection for the total output of the system than for the main product, wheat grain, because the secondary product, wheat straw, is totally unprotected (and thus has an NPCO of $4.42 / 4.42 = 1.00$).

mary factor costs to permit social valuation. All intermediate input costs are thus divided into tradable-input or factor cost categories.

An analyst searching for the sources of divergences in tradable-input markets finds that departures from world prices nearly always are caused by distorting policies rather than market failures. This situation is identical to that of divergences affecting outputs. Although one should always look carefully for the existence of market failures, in most empirical analyses product market failures (for both outputs and tradable inputs) are assumed to be nonexistent or unimportant. This assumption is made in the study summarized in Box 12.2.

Interpretation of the transfer effects of tradable-input price policies follows closely that of output price policies. If a government desires to raise domestic prices, it can restrict imports (if the product is imported), subsidize exports (if the country is a net exporter of the item), or tax all domestic consumption of the good. To reduce input costs, a government can subsidize importables, restrict exportables by imposing export taxes or quotas, or subsidize all domestic consumption of the input item.

Often governments decide to subsidize specific agricultural inputs, such as improved seeds or chemical fertilizers, in order to encourage greater use of these inputs and adoption of new technologies. In this respect, tradable-input price policy may have different goals and results from output price policy. Whereas output policy raises or lowers profits per ton for all systems, tradable-input policy can be designed to favor systems whose technologies use the subsidized inputs intensively.

Nominal protection coefficients on tradable inputs (NPCIs) can be calculated to permit comparisons among agricultural systems that produce dissimilar outputs. Calculations of NPCIs for single inputs and for the total of tradable inputs are contained in Box 12.2. These results are the opposite from those for the NPCOs, because both higher private prices of output and lower private costs of tradable inputs lead to greater private profits. Hence, the larger the NPCOs and the smaller the NPCIs, the greater the policy transfers to agricultural systems.

These separate influences of commodity price policies can be combined in an indicator called the effective protection coefficient (EPC), which is defined as $(A - B) / (E - F)$. This measure uses the same information as the NPCO (A and E) and the NPCI (B and F). It is a useful way to indicate the extent of incentives or disincentives that systems receive from product policies. The EPC concept is illustrated in Box 12.3. Its main limitation as an indicator of incentives is that it does not incorporate any effects of policies that influence factor prices. This omission means that EPC results should be interpreted as measures of the incen-

Box 12.2. Tradable-Input Transfers in a Portuguese Wheat System

	Tradable input costs (in escudos per kilogram)			
	Fertilizer (urea)	Spare parts (for repairs)	Other	Total
Private prices	1.35	1.93	6.25	9.53
Social prices	2.21	1.58	8.00	11.79
Effects of divergences (private prices less social prices)	−0.86	0.35	−1.75	−2.26

As in Box 12.1, the effects of divergences are the result of distorting policy only, not of market failures. A number of distorting policies caused the observed market (private) prices of tradable inputs to differ from comparable world prices. The government provided a subsidy on all sales of urea fertilizer, including that produced locally and that imported; this subsidy amounted to 0.86 escudos per kilogram, or 39 percent of the cif import price: (2.21 − 1.35) / 2.21 x 100 percent.

In contrast, the government levied an import tariff on tradable spare parts (used in making repairs), which increased the average domestic price for these inputs by 22 percent: (1.93 − 1.58) / 1.58 x 100 percent. The tariff on tradable imputs thus caused domestic producers of wheat to have to pay more for their spare parts than they would have without the tariff. This policy, therefore, created a negative transfer of 0.35.

Numerous other tradable inputs are aggregated in the column titled "Other." The most important of these inputs is compound fertilizer, nitrogen-phosphorus-potassium (NPK), which was subsidized to 38 percent of the cif import price. That subsidy accounted for most of the positive transfer on "other" tradable inputs.

The last column in the table shows that the wheat system enjoyed a total positive transfer of 2.26 escudos per kilogram on its tradable-input costs. If the government had not intervened, the wheat farmers would have had to pay 11.79 escudos per kilogram, but the actual policies permitted this cost to be reduced to 9.53. This total positive transfer of 2.26 resulted from the policy combination of subsidies on urea fertilizer of 0.86 and on other tradable inputs (mostly compound fertilizer) of 1.75 and of an import tariff on spare parts that created a negative transfer of (0.35). The signs for entries in the table are the opposite of those here because each input transfer is subtracted from the output transfer in the calculation of net transfers $(L = I − J − K)$.

The NPCI allows the analyst to contrast the effects of distorting policies on tradable-input costs in two or more agricultural systems that produce either identical or dissimilar tradable outputs. An NPCI equal to 1 indicates no transfer, an NPCI greater than 1 shows a negative transfer (because input costs are raised by policy), and an NPCI less than 1 denotes a positive transfer (since input costs are lowered by policy). In this example, the NPCI for urea fertilizer is 1.35 / 2.21 = 0.61, and that for other inputs is 6.25 / 8.00 = 0.78, both showing the effects of the subsidies. However, the NPCI for spare parts, 1.93 / 1.58 = 1.22, exceeds 1 because the price-raising import tariff created a negative transfer. The average NPCI for all tradable inputs is 9.53 / 11.79 = 0.81, which again points to the positive transfer from the entire set of policies affecting tradable inputs.

Box 12.3. Effective Protection Coefficient for a Portuguese Wheat System

| | Amounts (in escudos per kilogram) | |
	Revenues	Tradable-input costs
Private prices	27.42 (A)	9.53 (B)
Social prices	22.79 (E)	11.79 (F)
Effects of divergences	4.63 (I)	2.26 (J)

The EPC is the ratio of the difference between revenues and tradable-input costs in private prices to that in social prices. In PAM notation, EPC = (A − B) / (E − F). The numerator of EPC, A − B, is value added in private prices; the denominator, E − F, is value added in world prices. The ratio thus shows by how much policies in the product markets cause observed value added to differ from what it would be in the absence of commodity price policies.

EPC is an indicator of the net incentive or disincentive effect of all commodity policies affecting prices of tradable outputs and inputs. An EPC greater than 1 means that private profits are higher than they would be without commodity policies; the transfer from both output and tradable-input policies, taken together, is positive. An EPC less than 1 indicates the opposite result; the net effect of policies that alter prices in product markets is to reduce private profits, and the combined transfer effect is thus negative.

An EPC can be calculated for each agricultural system. For the wheat system of this example, it is (27.42 − 9.53 = 17.89) / (22.79 − 11.79 = 11.00) = 1.63. The interpretation of this result is that the net impact of government policy influencing product markets—that is, output price policy and tradable-input price policy—is to allow the wheat system depicted to have a value added in private prices 63 percent greater than the value added without policy transfers (as measured in world prices). The NPCO (A / E) of 1.20 indicates that policies caused output prices to be 20 percent higher than they would have been if world prices had been allowed to set domestic prices. The NPCI (B / F) on all tradable inputs of 0.81 showed that costs of tradable inputs were only 81 percent of what they would have been at world prices. The EPC is a single indicator that combines these two results by using the data from both. It is a useful measure of the combined effects of commodity price policies, but it does not account for any effects of policy in factor markets.

tive effects of commodity price policies but not as indicators of the total impact of policies that influence prices and costs.

Factor Transfers

Factor transfers, K, are defined as the difference between the costs of all factors of production (unskilled and skilled labor and capital) valued in actual market prices, C, and the social costs of these factors, G. One distinguishes between the inefficiency-causing effects of distorting policies affecting either output or factor markets and of market failures in factor markets.

The existence of factor market failures in developing countries is the rule rather than the exception. Analysts will usually assume that factor markets are going to be imperfect unless careful examination shows that the private factor prices are reasonable approximations of social prices. An illustration of factor transfer interpretation is given in Box 12.4.

Net Transfers

Net transfers, L, are output transfers (I) minus tradable input transfers (J) minus factor transfers (K). Because each of the components of the net transfer is defined as the effects of divergences between private and social valuations, L is the net difference between private profits (D) and social profits (H). This double accounting definition of L follows directly from PAM's two accounting identities.

The measure of net transfer, a principal result of the PAM approach, is illustrated in Box 12.5. The value of L shows the extent of inefficiency in an agricultural system. If market failures are a large source of the net transfer, this measure indicates how much long-term government effort (price policy, investment, and regulation) will be required eventually to permit the economy to operate efficiently. If, instead, most of the L is traced to distorting policies, the government can increase efficiency by reducing the degree of distortion—unless such changes will seriously impair the attainment of nonefficiency objectives. L is, therefore, a key input into policy analysis.

These measures of net transfer can be applied to a wide range of agricultural and nonagricultural systems. Comparisons can be made among different systems producing the same output, a variety of agricultural systems, and different sectors in the economy. However, L alone is not sufficient for such comparisons, because it is denominated in currency units per hectare, per ton or kilogram of the commodity produced. Once again, ratios are required so that the indicators will be free of specific units.

The profitability coefficient (PC), defined as $PC = D / H$, is a measure of the degree to which net transfers have caused private profits to exceed social profits. Because $D = (A - B - C)$ and $H = (E - F - G)$, the PC extends the effective protection coefficient—defined earlier as $(A - B) /$ $(E - F)$—to include factor transfers. PC is a more complete measure than EPC because it provides an indication of the total incentive effect

Box 12.4. Factor Transfers in a Portuguese Wheat System

	Factor costs (in escudos per kilogram)			
	Unskilled labor	Skilled labor	Capital	Total
Private prices	0.02	3.48	3.90	7.40
Social prices	0.02	2.82	5.13	7.97
Effects of divergences (private prices less social prices)	0.00	0.66	−1.23	−0.57

The effects of divergences in the factor markets are the result of both underlying market failures and distorting policies. Both of these distorting influences typically cause observed factor prices to diverge from their social valuations. Three primary factors were identified in the illustrated wheat system; but only two of them, skilled labor and capital, were important costs.

Unskilled labor was a minor cost element, amounting to only 0.02 escudos per kilogram in both private and social prices. The factor transfer for unskilled labor is thus 0. The private wage rate is taken as a reasonable indicator of the social price of unskilled labor because neither significant market failures nor distorting policies were identified after careful observation. Information about employment opportunities was widely available to potential searchers, and a considerable amount of seasonal and multiyear migration of unskilled laborers occurred. Government policies to have employees pay pension contributions and health insurance were largely unenforced and thus were ignored by unskilled labor in agriculture.

For skilled labor, market failures were also judged to be absent. Again, ample information and widespread migration of workers showed evidence of a well-functioning market for skilled labor. The wage rate paid by wheat farmers and millers exceeded the social wage rate for skilled laborers because of distorting government policy. Above the market wage, employers also had to pay a percentage of the wage as a tax to provide funds for employee health insurance and pensions (akin to social security in the United States). These policies caused private wages for skilled labor to be an estimated 23 percent higher than social wages—that is, the level that might have been expected without the policies. The result for the system was a negative factor transfer of (0.66) because the social price, 2.82, was raised by policy to a higher private price, 3.48.

The factor transfer for capital was in the opposite direction. The social opportunity cost of capital was estimated at 8 percent plus inflation for the country. The actual interest rates being paid by wheat farmers, which ranged between 2 and 6 percent plus inflation, were less than the estimated social rate. This divergence resulted from the market failure of an underdeveloped capital market, associated with insufficient numbers of financial institutions in rural areas; a government subsidy on agricultural credit for borrowers, usually larger farmers, who qualified for it; and a government policy to ration credit at controlled interest rates that were below market-clearing levels. As a result of these divergences, the private costs of capital, 3.90, were only 76 percent of their full social value, 5.13; the level of the positive factor transfer was 1.23.

The total factor transfer is found by summation of the amounts for the individual factors. In this example, the negative transfer of (0.66) resulting from the tax on skilled labor is more than offset by the positive transfer of 1.23 caused by the capital subsidizing policies. The net result is a small positive factor transfer of 0.57.

Box 12.5. Net Transfers, Profitability Coefficient, and Subsidy Ratios to Producers for a Portuguese Wheat System

	Tradable factor (in escudos per kilogram)			
	Revenues	Input costs	Costs	Profits
Private prices	27.42(A)	9.53(B)	7.40(C)	10.49(D)
Social prices	22.79(E)	11.79(F)	7.97(G)	3.03(H)
Effects of divergence	4.63(I)	−2.26(J)	−0.57(K)	7.46(L)

The net transfer, L, of 7.46 escudos per kilogram is the output transfer, 4.63, less the tradable input transfer, (2.26), less the factor transfer, (0.57). By definition, L = I − J − K. The net transfer is also the difference between private profits and social profits. Hence, L = D − H; and in the example, 7.46 = 10.49 − 3.03.

The net transfer is the sum of all divergences that cause private profits to differ from social profits. In the illustrated wheat system, all of the transfers, except part of the transfer from capital, were the result of distorting policy, not of market failures. All three categories of policy transfers were positive, indicating that the government was providing support to the wheat system in each instance. Because social profits, 3.03 escudos per kilogram, were positive, the system could have operated profitably without any policy transfers. These transfers, 7.46, raised the profits actually received by farmers and millers from 3.03 to 10.49.

The measure of net transfer, L, cannot be used for comparisons among systems producing unlike outputs. The ratio formed for this purpose is the profitability coefficient: PC = (A − B − C) / (E − F − G) = D / H. It shows the extent to which private profits exceed social profits. In the example, PC = 10.49 / 3.03 = 3.46. Policy transfers (and a capital market failure) have permitted private profits nearly 3.5 times greater than social profits.

The subsidy ratio to producers is SRP = L / E, the ratio of the net transfer to the social value of revenues. The purpose of this indicator is to show the level of transfers from divergences as a proportion of the undistorted value of the system revenues. If market failures are not an important component of the divergences, the SRP shows the extent to which a system's revenues have been increased or decreased because of policy. For the wheat example, the SRP is 7.46 / 22.79 = 0.33. This result means that divergences—almost entirely distorting policies in this example—have increased the gross revenues of the system by one-third. If, hypothetically, all policies on tradable inputs and factors were removed, the wheat system's NPCO would have to be increased from 1.20 to 1.33 to permit the system to maintain the same level of private profits.

of policies, including those influencing factor markets. An illustration of PC is also provided in Box 12.5.

A second ratio indicator, used to measure net transfers across dissimilar systems, is the subsidy ratio to producers (SRP), defined as L /E. It shows how large net transfers from divergences are in relation to the social revenues of the system. The smaller the SRP, the less distorted the agricultural system. The SRP, converted to a percentage, also shows the output tariff equivalent required to maintain existing private profits if all other policy distortions and market failures are eliminated. It thus indicates how much incentive or disincentive the system is receiving from all the effects of divergences. Box 12.5 illustrates the calculation and interpretation of the SRP ratio.

The Policy Analysis Matrix and Agricultural Planning

Good policy analysts know that one key ingredient of success in their profession is to stay ahead of the game. In most instances, policy-makers claim to need answers within periods of time that are too short to permit analysis to be done. "I need it done yesterday" is the common request. If unprepared, the policy analyst has to employ methods without proper reflection on their appropriateness, cut corners in gathering and cleaning data, and rush results into drafts without time for reflection and full interpretation. In contrast, a prepared policy analyst is fully aware that the process of decision making in government will often leave inadequate time for complete analysis. Preparation entails adopting methods that can be flexible (that is, carried out with varying degrees of completeness) and gathering essential data in advance on a regular basis. The key, therefore, is to choose a small number of flexible methods and to do basic data gathering and analysis ahead of requests for information.

The purpose here is not to suggest an ideal set of methods and analyses that might be appropriate for any agricultural planning agency; the division of policy responsibilities differs enough among countries to make such a task unworkable. Rather, the idea is to show how PAM analyses can form an integral part of three types of agricultural policy analysis—agricultural prices, public investment projects, and public agricultural research allocations. Policy-makers typically want to know how agricultural price policies affect farm incomes, where new public investments in agriculture should be made, or why public funds should be spent on one line of agricultural research instead of another. If a planning agency were assigned responsibility for all three policy areas, the PAM could assist that agency in setting its research agenda.

The PAM and Price Policy Analyses

Policies are enacted with the intent of bringing about change. But to measure change, one needs to know the existing situation and to understand something about how it has evolved during the recent past. For price policy analysis, PAMs fulfill the first of these needs. One purpose of PAMs is to show the extent to which policies and market failures have influenced the levels of revenues and costs facing producers in some recent base year. The PAM method is designed specifically to permit a clear demonstration to policy-makers of the effects of agricultural and macroeconomic policies.

For price policy analysis, the PAM demonstrates empirically the relationships among different policies and market failures that cause private prices to diverge from their social values. It allows calculation of competitiveness (private profits), and it shows how profits change as policies are altered. The accounting framework is a consistent means of tabulating information required for price policy analysis. The results need to be qualified to permit comparisons of the PAM's efficiency focus with nonefficiency objectives.

Ideally, one would like to construct PAMs for all main systems biannually over a fifteen-to-twenty-year period in order to trace the evolution of policy effects. For nearly all countries, this goal is unattainable because of data limitations. As a partial substitute, one can usually construct price policy graphs for up to two decades. These graphs are drawn separately, using annual data, for each main agricultural commodity and input. Each graph shows the domestic wholesale price of the commodity (or input), the comparable world price (cif import or fob export), and the domestic policy prices (floor price for producers and ceiling prices for consumers), if such exist. The graphs provide visual interpretations of the recent history of price policy and complement PAMs constructed for one or two recent years. Reasonably up-to-date PAMs and price policy graphs are thus two essential pieces of baseline information needed for price policy analysis. An illustration of a price policy graph, showing rice prices in Indonesia between 1974 and 1985, is presented in Box 12.6. An example of the PAM method used to undertake analysis of the projected impact of policy changes in agricultural system profits is summarized in Box 12.7.

Box 12.6. Price Policy Graph for Rice In Indonesia

A price policy graph is an illustrative device to permit easy visual com-parisons of year-to-year movements in three kinds of price series—world prices (cif import or fob export, adjusted to a domestic wholesale market level), domestic market prices (at both the wholesale and farm levels), and domestic policy prices (guaranteed floor price to producers and announced ceiling prices to consumers). Price policy graphs allow quick visual reviews of the patterns of price levels and price stability. One item of interest is the extent to which domestic prices are higher or lower than world prices because of price policy. For price stability, the issues are whether intrayear domestic prices have been successfully maintained between announced producer floor and consumer ceil-ing prices, because of trade and buffer stocking policy, and whether interyear domestic or world prices, both adjusted for inflation, have been more variable. Such historical graphs, when continuously updated, are excellent complements to PAMs.

The following figure describes rice prices in Indonesia between 1974 and 1985. The National Food Logistics Agency (BULOG) successfully implemented a buffer stock policy for rice. Through good management and well-designed and well-located warehouses, BULOG defends a paddy floor price to farmers by buying at the announced floor price. The success of the floor price is demonstrated in the price policy graph; the wholesale price in East Java (the main production and consumption region in Indonesia) only rarely and tem-porarily fell beneath the policy-determined floor price.

The graph also shows the annual and trend levels of Indonesian and compa-rable world prices of rice. In setting domestic rice price levels, Indonesian policy-makers have attempted for the most part to approximate the expected trend of world prices. Between 1973 and 1982, the trend domestic price on average was somewhat lower than the trend world price. This disincentive to production was countered with technology and investment policies and with substantial subsidies on fertilizer to induce adoption of fertilizer-intensive high yielding varieties of rice.

PAM and Investment Policy Analysis

If the planning agency has constructed PAMs for the country's major agricultural systems, these matrices can also provide results that aid in the process of determining the allocation of public investment in agri-culture. PAMs show the levels of efficiency (social profitability, or H) of each agricultural system studied. Calculation of domestic resource cost ratios (DRCs) allows the comparison of efficiency among systems that produce unlike outputs. These DRCs offer useful information to invest-ment planners.

Rupiah/kilogram

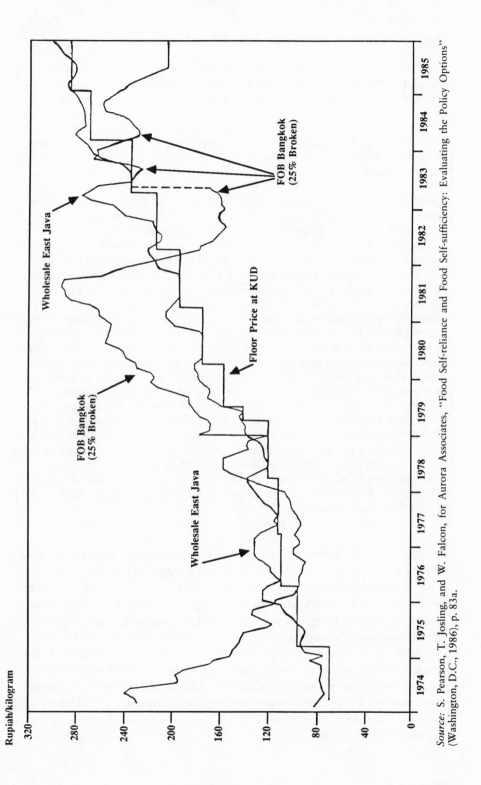

Source: S. Pearson, T. Josling, and W. Falcon, for Aurora Associates, "Food Self-reliance and Food Self-sufficiency: Evaluating the Policy Options" (Washington, D.C., 1986), p. 83a.

Box 12.7. The Projected Impact of Price Policy Changes on the Private
Profitability of Portuguese Agricultural Systems

The following table contains the results of private profitability calculations
for thirty-three Portuguese agricultural systems during the base year of data
collection, 1983, and projections for 1996. The set of agricultural prices that
faced producers in 1983 will undergo major changes because Portugal joined
the European Community in 1986. Moreover, until 1996, the country will
gradually align its agricultural prices to those of the Common Agricultural
Policy. The projected private profitabilities for 1996 thus reflect projections of
CAP prices and hence of Portuguese prices for that year.

Complete PAM analysis was carried out for all thirty-three systems, organized by commodity, region, and technology. But only the private profits are
reported in the table, because the policy question is whether adoption of the
CAP price regime will cause the need for large adjustments in any of Portugal's
agricultural regions. The projection results indicate that relatively easy adjustments are in store for the main farming systems in the center (the Ribatejo) and
in the good-soil areas in the south (the Alentejo); wheat and corn are projected
to become less profitable and sunflowers, sugar beets, tomatoes, melons, and
rice more profitable within the CAP regime. The private profits of dairying in
the Azores will decline but will remain positive, so no major difficulty is
foreseen there. Large losses in private profits are projected for the poor-soil
areas of the south (the Alentejo) and for the northwest. The large farms in the
south might need to convert their grain farms to pasture, forages, or forestry.
But the very-small-scale farmers in the densely populated northwest are likely
to experience a process of accelerated structural change if CAP prices cause
private profits to be as negative as those projected. In this way, construction of
PAM budgets for all of Portugal's principal commodity systems permits identification of whether large changes in price policy will likely trigger difficult or
easy regional adjustment.

Nearly all public investments in agriculture are made with the inten-
tion of reducing social costs in agricultural systems. (The exceptions are
those made to introduce new crops or technologies.) A critical element
in deciding on a strategy for a sequence of public investments is to know
the social profitabilities of the existing systems. Social benefits to public
investment are additions to positive social profits. Negative social profits could be reversed by removal of distorting policies. Hence, it is
critical for planners to know how socially profitable or unprofitable
systems are before the investment. PAMs provide this necessary baseline
information. They must be complemented with complete social benefit-
cost analyses of the most promising projects, selected on the basis of the
baseline social profits and expected improvements from the investments.

Farm-level profitability and land rent by soil type and crop, 1983 and 1996 (in thousands of escudos per hectare)

	1983 Profitability	1996 Profitability (base case)
The Alentejo		
Dryland, A and B soils:		
Wheat	23.0	1.1
Sunflowers	2.8	2.6
Dryland, C and D soils:		
Wheat	7.6	−8.0
Sheep, medium-technology	10.8	−1.3
Sheep, high-technology	3.8	−1.0
Beef, pasture-fed	2.4	−0.8
Irrigated:		
Rice	64.9	83.6
Tomatoes	79.3	85.1
The Ribatejo		
Dryland, sprinkler irrigation:		
Wheat	60.2	24.0
Corn	87.5	28.4
Sunflowers	31.9	33.8
Sugar beets	140.5	35.4
Wine		
Flood irrigated:		
Tomatoes	48.8	51.8
Melons	139.6	126.4
Rice	77.6	105.8
The Azores		
Dryland:		
Milk	36.0	24.4
The Northwest		
Dryland, traditional technologies:		
Milk	−85.4	−137.9
Corn	−0.5	−31.8
Potatoes	48.2	31.5
Wine	−43.4	−45.5
Dryland, medium technologies:		
Milk	75.9	−94.6
Corn	9.6	−34.3
Potatoes	61.4	48.4
Wine, ramada	27.7	19.0
Dryland, specialized technologies:		
Milk	56.5	−116.6
Potatoes	78.8	65.2
Wine, *cordao*	243.8	236.3

Source: Scott R. Pearson et al., *Portuguese Agriculture in Transition* (Ithaca: Cornell University Press, 1987), pp. 246–47.

Evaluations of alternative investment projects, therefore, can use the PAM baseline results to discover which systems are currently socially profitable and which are creatures of supportive policy. Project analysis consists of carefully altering certain costs or technical coefficients and comparing discounted time streams of costs and returns. The main caveat is that critical parameters—world prices, factor prices, and technologies—can change in the future; such changes must also be considered in project analysis.

PAM and Agricultural Research Policy Analysis

A similar situation arises in the analysis of public expenditures for agricultural research. Almost all such expenditures are intended to improve crop yields or to reduce input needs, thereby raising profits in existing agricultural systems. But it is not enough to know that the improved technology will reduce costs in a system. The key issue in choosing which system should receive attention is to know the relative social profitabilities of all of the systems for which technological improvements are possible. No social benefits accrue if technological change merely offsets existing negative social profit. Complementary analyses include projections of changes in world prices and factor prices along with technological changes arising from agricultural research, since the new technologies would be used in the future under differing economic environments.

The baseline PAMs show how well current systems are operating. The technological changes (yield increases or cost reductions) needed to arrive at improved private or social profits can then be determined relative to some starting point. Efficiency and nonefficiency objectives need to be evaluated separately, especially when potential technologies are developed for systems that begin with large negative social profits. An application of partial budgeting is described in Box 12.8; the example considers labor-saving technical changes in rice-farming systems in three West African countries—Burkina Faso, Mali, and Niger.

Communicating Results to Policy-Makers

Policy memoranda and oral reports are essential aspects of good policy analysis. If done effectively, they are the basis of the development of strong working relationships and mutual trust between economic technicians and policy-makers. Ultimately, economic analysis will be used importantly by policy-makers only if they are convinced that the analysis has been done correctly, has been based on all available information, and has been interpreted in ways that illuminate the choice they face. Effective communication, therefore, is a critical final step of policy analysis.

Some analysts are very good at the first three parts of policy analysis—understanding methods, collecting information, and interpreting results—but their effectiveness is limited because they are unsure how to explain the results to policy-makers. The inability to write a good policy memo is only rarely caused by the analyst's lack of skill in writing. Instead, it is often an inability to state information in ways that are easily understood by policy-makers.

Policy-makers as a group are busy people. Most have not studied economics at all (or lately), and some seem to believe that economics and economists exist more to cause problems for them than to help them make better-informed decisions. Only the few highly trained economists among them have any patience with technical economics jargon, and usually the few policy-makers who have been formally trained in economics are the only ones who receive much intellectual excitement from understanding the intricacies of economic methods. For many policy-makers, therefore, an inherent distrust of economics is combined with an intense dislike of economic jargon and methods. This common situation puts most economic analysts at a severe disadvantage. They must be able to communicate clearly, or they may be ignored.

Brevity and clarity in composing policy memos are aided by the use of consistent principles of organization. Busy policy-makers want to be sure that all relevant topics are covered in a logical order. For this reason, analysts are well advised to adopt a standard format to use in writing policy memos. One format for presenting the essential elements of policy memos is summarized in the seven numbered paragraphs below. The remainder of this section discusses each of the seven elements of this format. By following this organization for policy memos, analysts who have experienced difficulty in communicating with policy-makers should be able to improve the clarity and shorten the length of their memos. A series of short examples in the format is presented at the end of the section.

Essential Elements of Policy Memos

1. *Policy issues:* brief statement of (a) the specific policy issues to be addressed in the memo, (b) the aspects of the issues that the analysis covers, and (c) the

Box 12.8. Profitability and Technological Change in Rice Production in Three West African Countries

The table presents the results of partial budgeting analyses that investigated the social gain or loss from the introduction of alternative labor-saving technical changes in rice systems located in Burkina Faso, Mali, and Niger. The table was constructed with detailed information on several labor constraints, which appeared in the article from which the table is drawn. The results show the possibility of social gains from the introduction of animal traction, improved manual equipment, and small motorized threshers and the likelihood of social losses from the introduction of motorized techniques, which saved labor time but reduced labor productivity. This kind of analysis is also very informative for project planners or allocators of research funds, if the technical changes they might introduce would attempt to break labor constraints in the rice-farming systems. With relatively little effort beyond the initial construction and analysis of the budgets, the analyst can thus point out both baseline efficiencies and likely social gains or losses from specific technical changes.

wider policy context within which to view the specific policy under consideration.

2. *Method of analysis:* intuitive summary of (a) the basic logic of the method of analysis to be used; (b) why the method is appropriate for the particular policy question being studied; (c) how extensively the method has been applied in academic and policy analyses, locally and abroad; (d) the principal strengths and limitations of the method; and (e) the main qualifications that the method entails.

3. *Information needs:* summary listing of (a) the essential data requirements for the analysis, (b) complementary information that assists in the interpretation of results but is not essential for application of the method, (c) principal assumptions used for exogenous parameters or missing data, and (d) historical information used to provide a context for interpretation of the results.

4. *Interpretation of results:* full explanation of (a) the results obtained from analysis of the empirical information in the context of the selected method; (b) the sensitivity of the base-case results to changes in key data, parameters, or assumptions; (c) the meaning of the results within the selected method and within the context of the policy issue being studied; and (d) qualification of the results arising from limitations inherent in the method selected and from missing information.

5. *Implication of results for national interest groups:* brief summary of (a) the policy choices (usually to continue the status quo, do more, or do less), (b) the beneficiaries of successful research results, (c) the likely size of gains and losses for principal interest groups, (d) the main government objectives that would seem to be furthered or harmed by the policy choices, and (e) rough orders of

Net savings over manual cultivation from changes in techniques, inland countries* (in francs per hectare, except as noted)

Description	Labor saved (in days)	Value of labor saved[a]	Other indirect savings[b]	Additional direct costs of techniques	Other indirect costs[c]	Possible yield effects	Net gain
Basic manual system dam irrigation[d]	250	50,000	0	104,108	0	3.5	0
Ox land preparation and transport	36–41	7,800[e]	812[f]	5,264	112	Ambiguous	3,236
Power tillers	45	9,000	860[f]	14,410	576	Nil	−5,126[g]
Tractor plowing, seeding, and transport[h]:							
Compared to transplanting	95	19,000	1,760[i]	21,051[g]	2,697	Negative	−2,988
Compared to broadcasting	58	11,600	3,024[i]	22,209	521	Ambiguous	−8,106
Manual rotary hoe	12	2,400	48	223	0	Nil	2,225
Ox-drawn seeder and weeder:							
Compared to transplanting	55	11,000	720[f]	972	2,186[i]	Negative	8,562
Compared to broadcasting	20	4,000	2,140[f,k]	972	2	Positive	5,166
Herbicides	30	6,000	120	7,070	274	Nil	−1,224
Small motorized threshers:							
2.5 metric ton per hectare yield	23	4,500	0	2,120	0	Positive	2,380
3.5 metric ton per hectare yield	32	6,300	0	2,968	0	Positive	3,332
Large-scale stationary threshers[l]:							
Without transport	27	5,400	0	13,045	0	Negative	−7,645
With transport by tractor	37	7,400	0	17,122	0	Negative	−9,722

Source: Charles P. Humphreys and Scott R. Pearson, "Choice of Technique in Sahelian Rice Production," *Food Research Institute Studies* 17 (1979–1980): 254–55.

[a]At 200 francs per day.

[b]Includes estimated interest on working capital for labor and other inputs saved.

[c]Includes the estimated value of charges for working capital on expenses for operation and maintenance of new equipment and on other additional inputs.

[d]Values are totals per hectare, not incremental savings or costs.

[e]Based on thirty-nine labor days.

[f]Includes 500 francs saved because there is less use of hand tools.

[g]Assumes double cropping.

[h]Requires 35 horsepower tractor, disc plow, disc harrow, seed drill, and trailer.

[i]Includes 1,000 francs for hand tools.

[j]Includes 35 kilograms of extra seeds for drilling.

[k]Includes 25 kilograms of seeds saved by drilling.

[l]Assumes yields of 3.5 tons per hectare.

magnitude of the likely tradeoffs of government objectives associated with each of the policy choices.

6. *International ramification of results:* short discussion of (a) rough magnitude of the influence of policy choices on the country's quantities of import demands or export supplies of affected commodities, (b) likely impact of the policy choices on international flows of capital or labor, and (c) likely effect of the policy choices on the country's international diplomacy, including obligations to international organizations such as the World Bank, the International Monetary Fund, and the General Agreement on Tariffs and Trade.

7. *Summary of the pros and cons of policy changes:* single-paragraph summary that (a) highlights the lessons of the empirical analysis, (b) states clearly what the analysis contributes to the policy debate, (c) identifies the likely consequences for interested parties of each of the policy choices but does not offer any recommendations on selection among the policy choices.

Policy Issues

The first suggested element in the policy memo is a brief, clear statement of the specific policy issues addressed in the memo. This statement then should be both narrowed and broadened. It is narrowed by clarification of the exact aspects of the issue that can be addressed in the analysis, and it is broadened by the statement of how the specific issue fits into the wider policy context. The point is to be very clear about the limits of the analysis and about how the results fit into the bigger picture. This task is best done in one long or two short paragraphs of less than one page.

Method of Analysis

The next entry in the memo is an intuitive summary of the method of analysis that has been used to generate the results. This section is often the hardest one for analysts to write effectively because they tend to tell policy-makers more than they want or need to know. This part of the memo, above all others, must be clear and brief; otherwise, policy-makers will be forced to take the results on faith—since they will not have been able to understand how they were obtained—or to ignore the whole exercise.

How much to write depends in part on the complexity of the method. In general, however, the entire discussion of methods of analysis should not be more than one page. It should normally cover the five components outlined under the heading "Methods of Analysis" above. The first two are the most important. Even though the policy-maker probably is not interested in technical details, the basic logic of the method and why it is appropriate for the specific policy question being studied should be addressed. Stating these two things briefly can be difficult; teachers of economics often require several years before they understand methods well enough to explain them in simplified terms. Analysts new to a method thus might want to seek the assistance of those who have had more experience with it. The explanation needs to be made intuitive for policy-makers or it will fail.

The three other parts of summarizing the method are more straightforward. Policy-makers should be told whether the method is well known, fairly standard, or experimental; what strengths and weaknesses of the method will influence the results for the policy in question; and what qualifications are usually made to results obtained with the method. The discussion in this part should focus solely on method; it should not anticipate the results that will be reported later in the memo.

Information Needs

The section on information needs is perhaps the easiest to prepare, because it is rarely difficult for policy-makers to follow a discussion of information needs. There is sometimes a temptation, however, for analysts to offer excessive and lengthy detail. The rule, again, is to provide only as much as the policy-maker needs to know. But because the results from the analysis are necessarily only as good as the quality of the information used to generate them, policy-makers do need to know a lot of the detail concerning data inputs. This section, therefore, often runs to two pages.

It is helpful to divide information needs into four categories. The most critical category lists the essential data requirements for the analysis. In all economic methods, certain kinds of data are so important that they drive the system, since the results depend fundamentally on them. The second category assists in interpretation of the results but is not required for application of the method. If data in the first category are unavailable, the method cannot be used; if data in the second category cannot be found, the method can still be used, but some of the richness in interpretation of the results is lost. Policy-makers also need to hear briefly about a third kind of information—the main assumptions used for parameters that are entered from outside the method and the procedures used to substitute for missing data. Finally, it is desirable to provide policy-makers with historical information to help them place the results in a broader context. Often, they will already have this background information.

Interpretation of Results

Because the interpretation of results is the central part of the exercise, it is located at the center of the policy memo. Here is where the analyst has to explain what the results are and what they mean for the issues under study. This process can require up to two pages (or even more for larger studies).

Experience points to a four-step procedure in setting forth and explaining results of policy analysis. The first and most obvious step is to catalog the principal results obtained from analysis of the empirical information through use of the selected method. The trick is to scale down the mass of possible results and to report only those that are specifically used in the policy discussion. Usually, a second category of results comes from carrying out sensitivity analysis—that is, changing key data, parameters, or assumptions to study the effects on major results. A third and more difficult task is explaining the meaning of the results, first in the context of the method and then for the policy issue under examination. This task requires a focus on the results from the viewpoint of information and insights that policy-makers will need to make better decisions. The fourth kind of interpretation is qualification of the meaning of the results because of inherent limitations in the method or missing information. The purpose is to let policy-makers know how much faith they should have in the results.

Implications of the Results for National Interest Groups

The extension and summary of the results for national interest groups include several lessons that policy-makers typically require. Five steps are suggested: (1) reviewing the policy choices; (2) pointing out the likely gains and losses with each of the main choices; (3) making rough estimates, if possible, of the magnitude of the gains and losses for each of the principal interest groups; (4) identifying the primary government objectives (efficiency, income distribution, food security) that would be affected positively or negatively by the policy choices; and (5) sketching estimates, where feasible, of the size of the likely tradeoffs of government objectives associated with each of the policy choices. The purpose is to clarify the impact of policy change on political interest groups and on government objectives. It is not desirable for the analyst to include personal value judgments about good or bad outcomes. The task of the analyst is to make objective evaluations of the likely impacts of potential policies. The policy-makers then must choose among the outcomes.

International Ramifications of the Results

The section on international ramifications of the results is especially important for countries that are large traders on international markets and key actors in the international economy. It is less critical for small developing countries that are price-takers in the world markets and that generally follow rather than make international economic trends. Still, all countries need to be concerned about the international ramifications of their domestic policy actions.

Policy-makers need to be warned if domestic policies might have negative international effects. What is suggested here is a brief summary—only one paragraph unless the international effects are unusually large. The summary might contain references to three possible kinds of international influences: international trade effects and consequent impacts on world prices, if any; international factor effects (foreign investment and labor migration); and implications for international diplomatic obligations, including consistency with membership in international organizations and impacts on bilateral foreign policy.

Summary of the Pros and Cons of Policy Choices

The executive summary of pros and cons of policy choices should consist of a single paragraph aimed at exceptionally busy people in the highest ranks of government. It should state the essence of the policy memo. Like the body of the memo, it should not recommend policy choices. The summary should focus on three topics: (1) lessons of the empirical analysis—that is, the principal results; (2) contributions of the analysis to the policy debate for the specific issues being addressed; and (3) identification of the likely consequences of the policy choices for interested parties.

Illustration of Elements of a Policy Memo

1. *Policy Issues*
a. Our government is considering whether to allocate a substantial amount of agricultural research resources to the development of high-yielding wheat varieties for the good-soil areas of the southern region.
b. This memo summarizes the results of research measuring the degree of efficiency and the effects of government policy on the existing technology for producing wheat in the target zone.
c. These research results need to be complemented by similar analyses of the existing efficiency of other agricultural systems and of the potentials for cost-

reducing technological improvements in those systems so that the government can allocate its agricultural research resources most effectively.

2. *Method of Analysis*

a. The method of analysis used to measure the efficiency and effects of policy for the good-soil southern wheat system is the policy analysis matrix (PAM), which measures profitability in actual market (private) prices and in efficiency (social) prices.

b. The PAM method thus shows the actual revenues, costs, and profits that southern wheat farmers and millers are experiencing and those they would realize if they received sales revenues and paid the costs of production based on prices that would allocate resources most efficiently.

c. Variations of this method have been widely used in academic studies locally and abroad and in policy work in international aid agencies and agricultural research centers. However, this study is the first one based on the PAM in this ministry.

d. The principal strength of the PAM is that it gives measures of the economic efficiency of existing agricultural systems and of the effects of policy on those systems. Its main limitation is that its results are for a base year and thus need to be altered as principal parameters (such as world prices of outputs and inputs, wage rates, interest rates, and farming and processing technologies) change over time. The method, however, can readily accommodate such parameter changes.

e. The PAM efficiency measure, social profitability, is a requisite first step in the analysis. The next steps are to examine how much improved wheat technologies, developed with the research expenditure, might increase yields or save on inputs and thus reduce per unit costs and to contrast the results with those of similar studies for other systems that could benefit from more agricultural research.

3. *Information Needs*

a. The basic information required for PAM analysis is budget data (revenues and costs), broken down into prices and quantities for a representative wheat farm in the good-soil area of the southern region and for postfarm marketing and flour milling, world prices for products or inputs that are either imported or exported, and estimates of the efficiency values of wage and interest rates.

b. The basic PAM data need to be complemented by anticipated future changes in the budgets (related to the newly developed technologies), world prices, and factor (labor and capital) prices.

c. The budget data are complete and reliable, because they were compiled from agricultural census data, farm group information, and field surveys. The principal assumptions are that the social value of capital is 8 percent plus the rate of inflation and that the social value of skilled labor is 23 percent less than the actual market wage rate, reflecting taxes for pension contributions paid by employers.

d. No complete historical budget data for this area are known to exist. The current representative technology has spread gradually through the region during the past two decades.

4. *Interpretation of Results*

a. In the base year (1983), the representative wheat system was very profitable; private revenues were 27.42 (esudos per hectare) and private costs were 16.92; thus private profits were 10.50. Profitability was maintained at social prices. Social revenues, 22.79, were 4.63 less than private revenues because of import quotas on wheat; social costs, 19.76, were 2.84 above private costs mainly because of subsidies on fertilizers and credit; and therefore social profits, 3.03, although positive, were 7.47 less than private profits.

b. Projections to 1995 were made, using various assumptions about future world prices and factor costs, and the wheat system remained socially profitable under all reasonable sets of assumptions. No changes in technology were projected, because that analysis awaits information from agricultural research.

c. Two principal lessons emerge from these results. First, the current system operates efficiently, so all increases in social profit arising from new agricultural research will be net gains to the economy. Second, government policies—the import restrictions on wheat and the subsidies on fertilizer and credit—are resulting in excess private profits for good-soil wheat farmers.

d. The efficiency results appear robust because they are based on complete data and because they were realized under a wide variety of assumptions for key variables.

5. *Implications of Results for National Interest Groups*

a. The policy choice is whether the government should decide to allocate new research funds for southern region good-soil wheat.

b. The main beneficiaries of successful research results would be the wheat farmers and, to a lesser extent, the flour millers in the target region. The wheat farmers have farm wages and incomes that are currently among the highest in the country. They are already benefiting from agricultural price policies affecting wheat and inputs (see item 4). There are no obvious losers, other than taxpayers or those who would benefit if the research funds were spent elsewhere.

c. The size of the gains for wheat farmers is not yet estimable because no new budget data are now available on potential revenues and costs for the technologies to be developed with the research funds.

d. Successful research on wheat for the target area would likely advance two of the objectives of food policy but probably not the third. It would improve the efficiency of an already efficient system, and it would increase the productivity and reduce required imports for one of the country's staple foods, hence probably furthering food security. But the income distribution effects are not likely to be positive, because the technical innovations would aid mainly large, well-off farms that employ capital-intensive production technologies.

e. The policy tradeoff is thus a comparison of gains in efficiency and (probably) in food security with costs of income distribution. The decision will depend on the results of similar analyses for other commodities, technologies, and regions.

6. *International Ramification of Results*

a. Successful research is expected to reduce recent levels of imports of wheat by up to one-third, or a maximum of about 150,000 metric tons. This result is not

CHAPTER 13

The Practice of
Agricultural Policy Analysis

HOW MIGHT THOSE concerned with agricultural policy, as analysts
or policy-makers, conveniently approach the issues and organize their
research agendas? In particular, where does the policy analysis matrix
fit into the process of thinking about and measuring the effects of
agricultural policies? The purpose of this concluding chapter is to suggest answers to these two questions by summarizing the arguments
already presented. Two sections review the analytical approach to policy analysis. The first describes policies as instruments to achieve particular objectives. The second identifies when government intervention
can help an agricultural sector to run more efficiently and how an
analyst can approach the problem of measuring the effectiveness of
agricultural price policies. The scope of analysis is broadened in the
third section to include macroeconomic policies, especially exchange
rates, and linkages between those policies and agricultural price policies. The fourth section reintroduces the PAM approach as a way to
implement this analytical process and as an empirical method for measuring the effects of policy. A good complement to the PAM approach,
the construction of price policy graphs, is discussed briefly in the fifth
section. A final section then returns explicitly to the use of budgets as a
way to estimate PAMs and contrasts the strengths and limitations of
this method with the use of estimated elasticities to measure efficiency,
policy, and welfare effects.

Framework for Agricultural Policy Analysis

Governments are assumed to have broad objectives that they are
trying to further through interventions in the agricultural sector. The

255

three most common objectives are efficiency (the allocation of resources to effect maximal national output), income distribution (the allocation of the benefits of agricultural production to preferred groups or regions), and food security (the short-run stability of food prices at levels affordable to consumers, reflecting the adequacy of food supplies and the long-run guarantee of adequate human nutrition). Government actions that can further all three objectives are likely to be taken. Typically, however, the promotion of one objective conflicts with one or both of the others. In that situation, policy-makers must trade off gains in one area with losses in the others. For example, small loses in efficiency might be tolerated if the action were believed to result in significant improvements in income distribution or food security. Policy-makers make these tradeoffs explicitly or implicitly by forming value judgments about the worth of different objectives.

The need to make tradeoffs arises because of constraints in the economic system. Three categories of constraints limit the ability of policy-makers to realize all that they would like from their agricultural sectors. Production is limited by supply constraints—the input requirements of production technologies (for farming and processing) and the costs and availability of inputs. The value of the commodities produced is constrained in part by the characteristics of domestic demand—levels and growth rates of populations and incomes, changes in tastes and preferences, and willingness to substitute various agricultural commodities. Domestic supply and demand constraints are moderated by world prices for agricultural outputs and inputs. Because world prices, the third constraint, determine the domestic prices of internationally tradable commodities when no policies intervene, price policies either increase, decrease, or stabilize domestic prices relative to the underlying world prices. For each agricultural system, therefore, the three categories of constraints can be depicted by a drawing of a supply curve, a demand curve, and the relevant world price line for the outputs (the cif import price for goods that are partly imported or the fob export price for exported commodities).

Policies are the instruments of action that governments employ to effect change. Three principal categories of policies are used to bring about change in agriculture. The first is agricultural price policy. Two main types of price policy instruments can be used to alter prices of agricultural outputs or inputs. Quotas, tariffs, or subsidies on imports and quotas, taxes, or subsidies on exports directly decrease or increase amounts traded internationally and thus raise or lower domestic prices; these policies apply only to volumes traded internationally, not to domestic production. Domestic taxes or subsidies, in contrast, create transfers between the government treasury and domestic producers or consumers. Some cause a divergence between domestic and world prices; others do not.

The second category of policies is nationwide in coverage. Macroeconomic policy includes the central government's decisions to tax and spend (fiscal policy), to control the supply of money (monetary policy), and to impose macro price policies affecting the foreign-exchange rate (exchange-rate policy) and the domestic factors (wage, interest, and land rental rates). With the exception of land market policy, these decisions typically are not taken because of their impact on the agricultural sector. But macro policy effects, however unintended they might be, can more than offset the desired incentives of agricultural price policy.

In addition to price and macro policies, governments influence their agricultural sectors through public investment policy. Government budgetary resources can be invested in agriculture to increase productivity and reduce costs. The most common investments are in agricultural research to develop new technologies, in infrastructure (roads, irrigation, ports, marketing facilities), in specific agricultural projects to increase productive capacity and demonstrate new technologies, and in education and training of agriculturists to upgrade the human capital in the sector.

Effectiveness of Price Policies

The next step is to examine how the objectives-constraints-policies framework can be made operational. The analytical approach views policy-makers as enacting policies (price, macro, or investment) to further objectives (efficiency, distribution, or food security) in the face of economic constraints (supply, demand, and world prices). The main services policy analysis can provide to policy-makers are to distinguish whether a policy is likely to improve the efficient operation of the economy and thus raise the level of national income, to measure the expected magnitude of the efficiency gains or losses, and to quantify, when possible, the direction and extent of the policy's likely effects on the distributional and food security objectives. Even when the nonefficiency effects are difficult to measure, economic analysis can provide a reasonable estimate of the efficiency costs associated with the promotion of nonefficiency objectives.

The ways in which agricultural price policy can lead to efficient gains are limited to offsetting market failures, assisting agricultural infant industries, and stabilizing domestic prices. In developing economies, the most prevalent market failures usually are found in the factor markets, particularly for capital and occasionally for labor. These market failures are caused by insufficient development of institutions (such as financial intermediaries) and communication networks (so that information on jobs is not widespread). A second type of market failure is the existence of monopolies or monopsonies, where only one or a few (cooperating) sellers or buyers have the ability to manipulate market prices to their own advantage. Externalities (costs for which the person responsible cannot be charged or benefits that cannot be appropriated by the enterprise creating them) are a third source of market failures. Public goods are the principal source of externalities in developing countries. A public good is inadequately provided because not all of those benefiting from it can be charged for their use of it; governments thus invest in public infrastructure (roads, ports, large irrigation works), which would otherwise be inadequately supplied by private individuals.

The two other rationales for efficient intervention may also be viewed as responses to special kinds of market failures. One is to assist agricultural infant industries by correcting for dynamic market failures. The essence of the infant industry argument is that, over time, the existence of market failures (usually in the capital market or because of information bottlenecks) will cause insufficient investment and technical change and thus not permit the economy to benefit from dynamic learning effects. The presence of efficient operations in the future is not enough to justify policy that offsets the market failures; the efficiency cost to society of the inefficient use of resources in the early years must be compensated by larger efficiency gains in the later years.

The third rationale is to stabilize domestic agricultural prices (relative to unstable world prices) when insurance markets are absent. Governments perceive benefits from reducing price risks for producers, fending off consumer pressures, averting hunger if food crop prices rise, and avoiding adjustment costs for producers and consumers. Price stabilization requires public intervention in international trade and domestic marketing (transport and storage). If a public agency manages a buffer stock, the benefits from price stability may justify producer prices that are lower than average world prices and consumer prices that are higher than world prices. This margin should cover the costs of buffer stock management.

The circumstances for efficient intervention—offsetting of domestic market failures, assistance to infant industries, and stabilization of domestic prices—are potentially widespread. Analysts of efficient policy intervention look for the sources of market failure and assess the present and future benefits and costs of such policies. Even efficient intervention typically has costs as well as gains.

The analyst of nonefficiency objectives begins by measuring constraints and then considers the effects of policy on objectives. If full information is available on a single commodity, the analyst can summarize the supply constraints into a supply schedule and the demand constraints into a demand schedule and can draw a standard price-quantity diagram that will portray the situation before the policy is enacted.

If a policy to raise the price of a commodity—for example, a tariff on competing imports—is put into place, the analyst examines the hypothetical effects of the restrictive trade policy on government objectives. The tariff (tax on imports) raises the domestic price to producers and consumers and influences the quantities produced, consumed, and traded internationally. Facing a higher price, producers will increase output (because they can cover higher costs of production), consumers will cut back consumption (and shift to cheaper substitutes), and the country's demand for imports of the commodity will decline on both accounts. The impact on efficiency will be negative—producers will overproduce and consumers will underconsume relative to the world price—unless the higher domestic price serves to offset a market failure. The trade policy will redistribute income, causing transfers from consumers (who will consume less at the higher price) to producers (who will grow more at the higher price) and to the government treasury (which will receive the tariff revenue on remaining imports); the effect on distribution will depend on how well off the producers and consumers are without and with the policy. The influence of the policy on food security depends on the relative stability of the additional domestic output versus the imports it replaces.

This simplified example shows how a price policy can be analyzed in order to identify its effects on government objectives. In actual price policy analysis, the process is more complicated. The first step in choosing among policies is to investigate the feasibility of the policy instrument. The imposition of a tariff on imports of a commodity can be done readily if rampant smuggling can be prevented, whereas the distribution of subsidy payments to millions of small-scale farmers might not be feasible. The next step is to measure the administrative costs of implementing the feasible instruments—for example, the costs of hiring additional customs agents. Such costs should be added to any efficiency losses of the policy or subtracted from any gains. In this way, the analyst can incorporate policy feasibility and administrative costs.

Linkages between Macroeconomic and Agricultural Price Policies

The objectives-constraints-policies framework applies to macroeconomics policy as well as to price policy. Common macroeconomic objectives include rapid economic growth, a desirable distribution of national income, reasonably low unemployment, and moderate or low inflation. In addition to facing the sectoral constraints of supply, demand, and world prices, macroeconomic planners are also confronted by a need to maintain an approximate balance in the national fiscal accounts (government revenues and expenditures) and in the foreign-exchange accounts (export earnings and foreign capital inflows versus import expenditures and foreign capital outflows). The macroeconomic policies available to further these objectives in light of such constraints include fiscal and monetary policies, budgetary policies, and macro price policies influencing the foreign-exchange rate, interest rate, wage rate, and land rental rate.

The direct effects of macroeconomic policy on agricultural systems are felt through the macro price policies, especially exchange-rate policy. Fiscal and monetary policies influence agricultural systems indirectly by the interest and exchange rates. Budgetary policy—decisions on allocating both the recurrent and the capital budgets of the national government—also have indirect effects on systems, because budgetary choices influence agricultural price policy (through the availability of recurrent funds for subsidies) and public investment policy for agriculture. The three kinds of macro price policies affecting factor prices can be important in individual factor markets, although little can be said about them in general.

Some useful general lessons can be drawn from the relationships among fiscal and monetary policy, inflation, and the exchange rate and those between the exchange rate and price policies. Inflation is caused principally by macroeconomic policy—decisions to run fiscal deficits financed by expansionary monetary policy—abetted by inflation abroad that causes the prices of imports and exports to rise. If the government chooses to have a fixed-exchange-rate regime, the exchange rate will be changed only through discrete policy decisions, not because of market forces. When governments create inflation and then choose not to depreciate the nominal value of their currencies (by changing the exchange rate so that more units of domestic currency are required for each unit of foreign currency), profits are squeezed in agricultural systems that produce tradable commodities. The real exchange rate becomes overvalued when the rate of depreciation is less than the rate of inflation. Overvaluation of the real exchange rate imposes an implicit tax on producers of tradables (by keeping the domestic currency prices of their outputs artificially low), forces farmers growing tradable food crops to pay implicit food subsidies that benefit consumers, and permits artificially cheap imported inputs. A policy creating inflation with fixed nominal exchange rates squeezes agricultural profits, transfers the burden of subsidizing food from the government treasury to farmers, and makes projects based on tradable inputs appear to be more profitable than they would be if the exchange rate were set appropriately.

This state of affairs can be corrected if a government chooses to change the exchange rate. Devaluations are often difficult actions to take politically, because their short-run effects usually benefit rural inhabitants who have limited political power and harm powerful urban interest groups. Some form of foreign-exchange rationing is inevitable when the real exchange rate is overvalued, and this rationing is most often achieved by quantitive restrictions on imports that compete with domestically produced manufactures. Politically powerful urban manufacturers and their employees then shift from being supporters of devaluation to being vocal opponents of it. The prices of their products are protected from the taxing effects of overvaluation by the import quota, and the overvalued exchange rate permits them to obtain tradable inputs at artificially low prices.

The Policy Analysis Matrix

A central theme of this book is that the PAM approach to agricultural policy analysis can provide decision-makers and analysts with both a helpful conceptual construct for understanding the effects of policy and a useful technique for measuring the magnitudes of policy transfers. Because the accounting matrix is simultaneously a teaching tool and a way of undertaking and reporting empirical analysis, PAM results can be communicated easily to policy-makers, who might not be specialists in economics.

Three related questions can be addressed with the PAM approach. Ministries of agriculture are concerned with the competitiveness of their countries' principal farming systems; actual income received by farmers is thus the first issue examined with the PAM method. Ministries of economic planning focus on the growth and distribution of national income, and planning agencies of agricultural ministries want to maximize agricultural income; the efficient allocation of resources in agriculture (and elsewhere in the domestic economy) is therefore the second issue addressed by the PAM. Decision-makers throughout the government—including those acting on agricultural price policy, others concerned with macroeconomic policy, and yet others dealing with the allocation of public investment to the agricultural sector—want to be informed about the effects of policy and of market failures. Each policy analysis matrix is thus constructed to address these three central issues of agricultural policy—competitiveness, efficiency, and policy transfers.

For PAM analysis to be carried out, an accounting matrix is constructed for each representative agricultural commodity system. An agricultural commodity system consists of a farm technology for producing a commodity (or set of commodities) in a given agroclimatic zone, a way of moving the crop from the farm to a processing site, a technology for processing the crop into marketable products, and a way of transporting the products to wholesale markets. Because all farms differ somewhat from one another, some aggregation needs to be done so that the empirical analysis becomes manageable. The identification of representative agricultural systems reflects differing aggregate combinations of commodities produced, technologies used, and agroclimatic locations of production. A study of one staple food commodity in a country might identify few or many representative systems for that commodity, depending on the complexity of technologies and agroclimatic conditions.

Each matrix is a combination of two accounting identities, one defining the rows and the other the columns. The first identity is the profits identity: revenues less costs equal profits. The second identity is a definitional statement of efficiency, or social valuations of revenues, costs, or profits. Actual market, or private, valuations of these entries are observed by surveying analysts. These private observations can diverge from the underlying social valuations for one of two reasons. The first source of divergence between private and social valuations is the category of market failures—factor market imperfections, monopolies or monopsonies, and externalities, including public goods. Any of these failures of markets to work efficiently can cause inefficient pricing signals. The second and more widespread source of divergence is the existence of distorting government policies. As noted earlier, efficient policies offset market failures; all other policies distort the economy, moving it away from its most efficient allocation of inputs and outputs.

Distorting policies are not necessarily inappropriate; they can be justified if their efficiency losses are more than offset by gains from the furthering of nonefficiency objectives. The two sources of divergences—market failures and distorting policies—cause private prices to differ from social prices of revenues, costs, and profits. The definitional identity for each column of a PAM is therefore known as the "effects of divergences" identity: private prices less social prices equal the effects of divergences.

The empirical estimation of PAMs proceeds from these two identities. Two fundamental steps are involved in preparing the research inputs into a PAM. The first is building budgets in private prices for the representative systems. To complete this step, the analyst compiles existing information on farm management studies and verifies and completes the farm budget data through field surveys. The farm budgets are then complemented with postfarm budget data on transporting and processing. This private budget information on revenues and costs is entered into the first row of PAM. Use of the profits identity allows calculation of private profits or competitiveness, the first research output of the PAM analysis.

The second step in building a PAM is to convert the entries for revenues and costs in private (actual market) prices into counterpart entries in social (efficiency) prices. The calculation of social prices is a combination of science, art, and guesswork, as all practitioners of social benefit-cost analysis are well aware. The approach followed in this book has been to explain fully why some dimensions of social valuations are extraordinarily complicated to handle empirically and then to suggest shortcuts that usually work well. The social valuations of outputs and inputs that would enter into international trade (in the absence of distorting trade policy) are given by their comparable world prices (cif import prices for importables and fob export prices for exportables). World prices, even if set in less than fully competitive international markets, provide a valuation standard of the choice the country has to use world markets or not. In the absence of distorting trade policy, the world prices determine the domestic prices of tradables and create efficient allocation.

Social valuation of inputs that do not enter into international trade is more difficult on both conceptual and empirical grounds. Most problematic are the social prices of the primary factors of production—labor, capital, and land. In principle, the observed, private factor prices have to be corrected for the distorting influences of divergences in output markets, market failures in factor markets, and distorting government policies in factor markets—in short, for all divergences in the economy. This practically impossible task is therefore roughly approximated with a series of rules of thumb meant to guide the analyst in careful observation of key factor markets and policies. The other kind of inputs that are nontradable internationally are some intermediate inputs into farming, marketing, and processing. The nontradable inputs, such as electrical power and truck transportation, are disaggregated into their component costs of tradable inputs and primary factors. These indirect costs are then added to the direct costs of tradables and factors used in the system. For this reason, each PAM has only two cost column categories—tradable inputs and primary domestic factors.

The second research output from PAM analysis, the calculation of social profits or efficiency, follows easily from application of the profits identity—once the analyst has found social valuations for revenues (the world prices of outputs), tradable input costs (their world prices), and factor costs (their social opportunity costs, or the amounts of national income forgone from their not having been used in their best alternative occupations). Positive social profit is a measure of efficiency, or comparative advantage, because the value of the goods produced by the agricultural system exceeds the costs of production after all causes of inefficiency—distorting policies and market failures—have been (hypothetically) removed. Negative social profit indicates the opposite result; the country is wasting resources by allowing inefficient production, which occurs because of distorting policies (which might be serving other government objectives) or market failures (which the government is unable or unwilling to correct with efficient policy).

The third row of each PAM, which measures the effects of divergences, is determined by application of the second definitional identity: private prices less social prices equal the effects of divergences. Occasionally, an analyst has better information on a third row entry than on its second row counterpart; thus social valuation is an output of rather than an input into the analysis. Typically, however, the divergences are research outputs. The analyst's job is not always completed at this point. Sometimes policy-makers need to have the effects of divergences broken down into those associated with market failures and those caused by particular policies. For the product markets (in which private prices of tradable outputs and inputs are determined), the analyst should try to identify market failures; if none are found, product market failures can be assumed to be nonexistent, unimportant, or unmeasurable. For the factor markets, the opposite expectation is held; and divergences that cannot be associated with distortions in the output or factor markets are assumed to be the result of factor market imperfections. The measured divergences or transfers for outputs and tradable inputs will generally be the result of distorting policy, whereas those for factors will be caused by a combination of distorting policy and factor market imperfections.

The close linkages between exchange-rate policy and price policy are also observed readily in the PAM. When distorting policies cause private product prices to diverge from their social values under an appropriate exchange rate, all of the measured transfer in the third row of a PAM is caused by price policies. But when the exchange rate is overvalued, the social valuations of both tradable outputs and tradable inputs need to be adjusted to reflect the degree of overvaluation; for example, a 20 percent overvaluation would need to be corrected by a 20 percent increase in the amounts for social revenues, social input costs, and social profits. The third row would show exchange-rate policies taxing output revenues, subsidizing input costs, and taxing profits.

The construction of a PAM, therefore, normally entails the finding of information on private revenues, private tradable input costs, private factor costs, social revenues, social tradable input costs, and social factor costs. Application of the profits identity yields two research outputs—private profits (competitiveness) and social profits (efficiency). The four other research outputs—output transfers, tradable-input transfers, factor transfers, and net policy transfers—are found through use of the divergence identity. The net transfer—the difference between private and social profits or the combination of all three other kinds of transfers—results from the complete set of agricultural price and macroeconomic policies and market failures that influence the system.

Because the data for the PAM represent a chosen base year, the results are static and potentially applicable to only that year. Projections of changing future world prices, technologies, and factor prices can be made to simulate paths of dynamic comparative advantage, as social profits change in response to varying parameters. Investment policy analysis can be assisted by the construction of baseline PAMs, identifying social profits before any public investment, and by analyses of dynamic comparative advantage with and without the prospective investment. The PAM approach can thus be used to illuminate baseline conditions and then to measure the effects of changing price, macroeconomic, or investment policies on the private and social profits of agricultural systems in the base year or in the future as key parameters change.

Price Policy Graphs

A set of PAMs for the country's principal representative agricultural systems provides analysts and polcy-makers with informative pictures of the existing structure of policies affecting agriculture and with a useful analytic tool for investigating the effects of future policy change. However, in most countries, there is no information base to permit construction of historical PAMs that would show changes every two or three years as trends in world or factor prices and technologies changed. Budget data might be available at best for a few systems during scattered years. But informed policy analysis requires an understanding of the recent history of policy changes as well as the detailed array of profitabilities in a given base year. This need can be met at least partially by the construction of price policy graphs.

A price policy graph is a device to permit easy visual comparisons of year-to-year movements in three price series—world prices (cif import or fob export, adjusted to a domestic wholesale market level), domestic market prices (at both the wholesale and farm levels), and domestic policy prices (guaranteed floor prices to producers and announced ceiling prices to consumers). Price policy graphs, based on annual data for fifteen to twenty years in the recent past, can be constructed for the principal agricultural commodities produced and for the main tradable inputs into agriculture. They allow a quick visual review of the pattern of price levels and price stability. If historic price policy graphs are continuously updated, they can serve as particularly useful complements to PAMs in the presentation of policy analysis.

Concluding Comments

Several practical lessons for practitioners emerge from this study of agricultural policy analysis. Approaches to issues and the policy agenda can be organized within the objectives-constraints-policies framework, and diagrammatic analysis can be used to identify the general direction of policy effects. Historical perspective can be provided through a compilation of price policy graphs for the most important agricultural products and inputs. Much insight is gained from using the PAM approach to the quantitive analysis of agricultural systems. The construction of PAMs, complemented by historical price graphs, provides essential baseline information for the analysis of agricultural policy.

The standard approach to agricultural policy analysis relies on estimated elasticities of supply and demand. When policies raise or lower market prices, use of the elasticities permits the analyst to quantify changes in amounts produced and consumed; income transfers among producers, consumers, and the government treasury; and efficiency losses or gains. The PAM calculations usually are based on budget data, not elasticities. A strength of the PAM method is the disaggregation of supply in terms of technology and agroclimatic zone. Such disaggregation permits a detailed understanding of constraints on systems and provides a basis for the analysis of investment and technological change influencing the dynamic comparative advantage of agricultural systems. The principal weakness of the PAM approach is that empirical applications may not correctly specify all the marginal adjustments to alterations in output and input prices. Without sufficient information (such as elasticities of output supply and input demand), exact PAMs cannot be constructed, and approximations must be made. Unless this is done, the empirical researcher will be left with nothing more than a numberless diagram, little understanding of how the many divergences affecting agricultural systems offset one another, and no input into the policy-making process. Budget-based PAMs fill this gap in agricultural policy analysis.

Index

Library of Congress Cataloging-in-Publication Data

Monke, Eric A.
  The policy analysis matrix for agricultural development.

  Bibliography: p.
  Includes index.
  1. Agriculture and state—Decision making—Econometric models. 2. Agriculture and
state—Developing countries—Decision making—Econometric models. I. Pearson, Scott R.
II. Title.
HD1415.M536   1989   338.1'8'091724   88–47938
ISBN 0–8014–1953–0 (alk. paper)
ISBN 0–8014–9551–2 (pbk. : alk. paper)